Applied Mathematical Sciences

Volume 70

Applied Mathematical Sciences

1. *John:* Partial Differential Equations, 4th ed.
2. *Sirovich:* Techniques of Asymptotic Analysis.
3. *Hale:* Theory of Functional Differential Equations, 2nd ed.
4. *Percus:* Combinatorial Methods.
5. *von Mises/Friedrichs:* Fluid Dynamics.
6. *Freiberger/Grenander:* A Short Course in Computational Probability and Statistics.
7. *Pipkin:* Lectures on Viscoelasticity Theory.
9. *Friedrichs:* Spectral Theory of Operators in Hilbert Space.
11. *Wolovich:* Linear Multivariable Systems.
12. *Berkovitz:* Optimal Control Theory.
13. *Bluman/Cole:* Similarity Methods for Differential Equations.
14. *Yoshizawa:* Stability Theory and the Existence of Periodic Solution and Almost Periodic Solutions.
15. *Braun:* Differential Equations and Their Applications, 3rd ed.
16. *Lefschetz:* Applications of Algebraic Topology.
17. *Collatz/Wetterling:* Optimization Problems.
18. *Grenander:* Pattern Synthesis: Lectures in Pattern Theory, Vol I.
20. *Driver:* Ordinary and Delay Differential Equations.
21. *Courant/Friedrichs:* Supersonic Flow and Shock Waves.
22. *Rouche/Habets/Laloy:* Stability Theory by Liapunov's Direct Method.
23. *Lamperti:* Stochastic Processes: A Survey of the Mathematical Theory.
24. *Grenander:* Pattern Analysis: Lectures in Pattern Theory, Vol. II.
25. *Davies:* Integral Transforms and Their Applications, 2nd ed.
26. *Kushner/Clark:* Stochastic Approximation Methods for Constrained and Unconstrained Systems
27. *de Boor:* A Practical Guide to Splines.
28. *Keilson:* Markov Chain Models—Rarity and Exponentiality.
29. *de Veubeke:* A Course in Elasticity.
30. *Sniatycki:* Geometric Quantization and Quantum Mechanics.
31. *Reid:* Sturmian Theory for Ordinary Differential Equations.
32. *Meis/Markowitz:* Numerical Solution of Partial Differential Equations.
33. *Grenander:* Regular Structures: Lectures in Pattern Theory, Vol. III.
34. *Kevorkian/Cole:* Perturbation methods in Applied Mathematics.
35. *Carr:* Applications of Centre Manifold Theory.
36. *Bengtsson/Ghil/Källén:* Dynamic Meteorology: Data Assimilation Methods.
37. *Saperstone:* Semidynamical Systems in Infinite Dimensional Spaces.
38. *Lichtenberg/Lieberman:* Regular and Stochastic Motion.
39. *Piccini/Stampacchia/Vidossich:* Ordinary Differential Equations in R^n.
40. *Naylor/Sell:* Linear Operator Theory in Engineering and Science.
41. *Sparrow:* The Lorenz Equations: Bifurcations, Chaos, and Strange Attractors.
42. *Guckenheimer/Holmes:* Nonlinear Oscillations, Dynamical Systems and Bifurcations of Vector Fields.
43. *Ockendon/Tayler:* Inviscid Fluid Flows.
44. *Pazy:* Semigroups of Linear Operators and Applications to Partial Differential Equations.
45. *Glashoff/Gustafson:* Linear Optimization and Approximation: An Introduction to the Theoretical Analysis and Numerical Treatment of Semi-Infinite Programs.
46. *Wilcox:* Scattering Theory for Diffraction Gratings.
47. *Hale et al.:* An Introduction to Infinite Dimensional Dynamical Systems—Geometric Theory.
48. *Murray:* Asymptotic Analysis.
49. *Ladyzhenskaya:* The Boundary-Value Problems of Mathematical Physics.
50. *Wilcox:* Sound Propagation in Stratified Fluids.
51. *Golubitsky/Schaeffer:* Bifurcation and Groups in Bifurcation Theory, Vol. I.
52. *Chipot:* Variational Inequalities and Flow in Porous Media.
53. *Majda:* Compressible Fluid Flow and Systems of Conservation Laws in Several Space Variables.
54. *Wasow:* Linear Turning Point Theory.

(continued following index)

Applied mathematical sciences.

P. Constantin C. Foias B. Nicolaenko
R. Temam

Integral Manifolds and Inertial Manifolds for Dissipative Partial Differential Equations

Springer-Verlag
New York Berlin Heidelberg
London Paris Tokyo

P. Constantin
Department of Mathematics
University of Chicago
Chicago, IL 60637
U.S.A.

C. Foias
Department of Mathematics
Indiana University
Bloomington, IN 47405
U.S.A.

B. Nicolaenko
Center for Nonlinear Studies
Los Alamos National Laboratory
Los Alamos, NM 87545
U.S.A.

R. Temam
Department de Mathematiques
Université de Paris-Sud
91405 Orsay
France

Mathematics Subject Classification (1980): 58GXX

Library of Congress Cataloging-in-Publication Data
Integral manifolds and inertial manifolds for dissipative partial
differential equations / P. Constantin ... [et al.].
p. cm. — (Applied mathematical sciences; v. 70)
Bibliography: p.
Includes index.
ISBN 0-387-96729-X
1. Differential equations, Partial. 2. Manifolds (Mathematics)
I. Constantin, P. (Peter), 1951– . II. Series: Applied
mathematical sciences (Springer-Verlag New York Inc.); v. 70.
QA1.A647 vol. 70
[QA377]
510 s—dc19
[515.3′53] 88-20021

Typeset by Asco Trade Typesetting Ltd., Hong Kong.
Printed and bound by R. R. Donnelley & Sons, Harrisonburg, Virginia.
Printed in the United States of America.

9 8 7 6 5 4 3 2 1

ISBN 0-387-96729-X Springer-Verlag New York Berlin Heidelberg
ISBN 3-540-96729-X Springer-Verlag Berlin Heidelberg New York

Preface

This work was initiated in the summer of 1985 while all of the authors were at the Center of Nonlinear Studies of the Los Alamos National Laboratory; it was then continued and polished while the authors were at Indiana University, at the University of Paris-Sud (Orsay), and again at Los Alamos in 1986 and 1987.

Our aim was to present a direct geometric approach in the theory of inertial manifolds (global analogs of the unstable-center manifolds) for dissipative partial differential equations. This approach, based on Cauchy integral manifolds for which the solutions of the partial differential equations are the generating characteristic curves, has the advantage that it provides a sound basis for numerical Galerkin schemes obtained by approximating the inertial manifold.

The work is self-contained and the prerequisites are at the level of a graduate student. The theoretical part of the work is developed in Chapters 2–14, while in Chapters 15–19 we apply the theory to several remarkable partial differential equations.

We wish to thank G. R. Sell, J. M. Ghidaglia, M. Jolly, J. C. Saut, and E. Titi for their interest and remarks. We are also indebted to Fred Flowers for the careful typing of the manuscript.

Acknowledgments

This research was partially supported by the Applied Mathematical Sciences Program of the U.S. Department of Energy, Contract DE-ACd02-82ER12049 and Grant DE-FG02-86ER25020; by the National Science Foundation, Grant NSF-DMS-8602031, and the Research Fund of Indiana University; and by the Center for Nonlinear Studies, Los Alamos National Laboratory, operated by the University of California under Contract W-7405-ENG-36. P. Constantin acknowledges a Sloan research fellowship.

Contents

Introduction

Recently, considerable theoretical and computational evidence has accumulated supporting the remarkable similarities between the long-time evolution of solutions of dissipative partial differential equations (PDEs) and solutions of finite-dimensional dynamical systems, or ordinary differential equations (ODEs). For the latter, numerous studies have discovered and analyzed complex dynamical bifurcations of finite vector fields [CoE, De, Sch, GH, MeP, ChH, BPV]. Computer simulations for the dynamics of many dissipative PDEs evidence an equally rich complexity [HN1, HN2, HNZ, BLMcLO]. The connection between the long-time behavior of finite differential systems and that of PDEs was first established by the discovery that dissipative PDEs possess a finite number of asymptotic degrees of freedom: they have a compact, universal attractor X with finite Hausdorff and fractal dimension (modulo some regularity conditions) [BV, BV1, CF1, CFT, DO, He, Hl, HMO, MeP, MP, NST, NST1, T]. Estimates on the number of such degrees of freedom have been obtained for two- and three-dimensional turbulent continuum flows [CF1, CFT, CFMT]. Still, such results do not imply that, for a given dissipative PDE, the asymptotic behavior and in particular the universal attractor X coincide with those of an appropriate differential equation. Recently, it has been shown that for certain dissipative PDEs this is indeed the case. The equations possess finite-dimensional *inertial* manifolds, i.e., invariant manifolds that contain the attractor X and attract exponentially all trajectories (cf. [FST, FNST]). The restriction of the PDE to the inertial manifold is an ODE, which we call an *inertial form* for the given PDE; it is a global analog of the normal form from the Center Manifold Theory [Ha, Ca, GH, He]. The long-time behavior of solutions of a PDE possessing a large enough inertial manifold is completely determined by the inertial form.

Up to now all existence proofs for inertial manifolds have exploited the existence of suitably large spectral gaps for the principal linear partial differential operator in the dissipative PDE or system of PDEs* (see [FST, FNST, M-PS]). Therefore the existence inertial manifolds is still unresolved for several important equations, for instance the Navier–Stokes equations in space dimension 2.

This work, the main results of which were announced in [CFNT], focuses on a new geometric explicit construction of inertial manifolds from *integral* manifolds generated by some initial finite-dimensional surface. The method covers a large class of dissipative PDEs (see Chapters 15 to 19 below). The existence of a *smooth integral* manifold, the closure of which is an *inertial* manifold M (i.e., containing X and uniformly exponentially attracting), requires a more detailed analysis of the geometric properties of the infinite-dimensional flow. The method is explicitly constructive, integrating forward in time and avoiding any fixed-point theorems. This is in contrast to the techniques in the original works [FST, FST1, FNST, FNST1], where the construction of an inertial manifold is that of a global center manifold through backward–forward time integration.

The integral inertial manifold construction is well suited to the fast and robust numerical approximation of such manifolds.

The key geometric property upon which we base the construction of our integral inertial manifold M is a spectral blocking property of the flow, which controls the evolution of the position of surface elements relative to the fixed reference frame associated to the linear principal part of the PDE. (There is a deep connection between the spectral blocking property and the Sacker–Sell spectral theory for time-dependent linear systems of ODEs (cf. [SS]), which will be studied elsewhere.)

In Chapters 2 to 19 we present a general and flexible method for the construction of our inertial manifolds; we obtain an existence theorem that requires some conditions to be satisfied. Each of these must be specifically verified in applications; some involve inequalities for spectral gaps of the principal linear part; some are inequalities that ensure regularity properties of the equations as well as geometric and spectral properties of the infinite-dimensional flow and of the initial surfaces.

Under general conditions, we establish in Chapters 11 to 14 further properties of the inertial manifolds constructed by our method: they entirely control the dynamics of the infinite-dimensional flow (asymptotic completeness) and are robust with respect to Galerkin approximations and other perturbations.

In Chapters 15 to 19 we show the flexibility of our method by detailing the construction of inertial manifolds for several specific examples. In Chapter 15 we follow the theory presented in Chapters 2 to 10 to the letter for the

* Since the completion of this work, considerable progress was made in this respect for reaction–diffusion equations [MPS, HS].

Kuramoto–Sivashinsky equation. The nonlocal Burgers equation is considered in Chapter 16 and requires some minor adjustments. In both cases, the verification of the conditions of the general theorem produces somewhat large estimates of the dimension of the inertial manifold. This may be an artifact due to our proof; the construction may work for lower dimensions. Indeed, one can choose initial surfaces for some subclass of nonlocal Burgers equations which generate inertial integral topological manifolds (not necessarily smooth); these manifolds have all other properties although their dimension can be far lower than the estimates given by the sufficient conditions. The Cahn–Hilliard equation and reaction–diffusion equations and systems (including Chaffe–Infante's) are considered in Chapters 17, and 18, and 19. The implementation of the method of integral manifolds for these examples requires natural modifications of details of the construction. The method is highly flexible and is readily adapted to the special structure of each equation.

The next chapter contains a more detailed description of our results.

CHAPTER 1

Presentation of the Approach and of the Main Results

The long-time behavior of dissipative partial differential systems is characterized by the presence of a universal attractor X toward which all trajectories converge. This is the largest bounded set in the phase space of the system on which the backward-in-time initial value problem has bounded solutions. The structure of X may be very complicated even in the case of simple ordinary differential equations: X may be a fractal or parafractal set (i.e., a compact set for which the Hausdorff and fractal dimensions are different). In the case of dissipative partial differential equations, although the phase space (in the function space) is an infinite-dimensional Hilbert space, X has finite fractal dimension (see [CF, CFT]). However, the already complex nature of X is in this case further complicated by the infinite degrees of freedom of the ambient space.

We treat here a class of dissipative systems that possess inertial manifolds $\bar{\Sigma}$. These are positively invariant regular finite-dimensional objects toward which all solutions tend with (at least) a uniform exponential rate. Let H be the Hilbert phase space (usually a subspace of a Sobolev space) and let $S(t)u_0$ denote the trajectory (solution of the system) starting at $t = 0$ from u_0. By an inertial manifold for $S(t)$ we mean a set $\bar{\Sigma}$ satisfying

$\bar{\Sigma}$ is a finite-dimensional Lipschitz manifold, (1.1)

$S(t)\bar{\Sigma} \subset \bar{\Sigma}$ for $t \geq 0$, (1.2)

There exists a constant k such that, for every $u_0 \in H$, there exists $t_0 \geq 0$ and a constant $c > 0$ (uniform for u_0 in bounded sets) such that, for $t \geq t_0$,

$$\operatorname{dist}(S(t)u_0, \bar{\Sigma}) \leq c \cdot \exp(-kt). \tag{1.3}$$

We shall present here a geometric method of constructing $\bar{\Sigma}$ for a class of dissipative systems large enough to contain the one-dimensional Kuramoto–Sivashinsky, Cahn–Hilliard, and nonlocal Burgers equations and the one- and two-dimensional parabolic reaction–diffusion equations. Inertial manifolds for the Kuramoto–Sivashinsky equations were obtained in [FST1] and studied in [FNST1]. For the moment we are unable to treat the 2D Navier–Stokes equations, but a great part of our program goes through, and we expect the method, suitably modified, to work in this case also.

We shall treat equations

$$\frac{du}{dt} + N(u) = 0 \quad \text{with} \tag{1.4}$$

$$N(u) = Au + R(u), \tag{1.5}$$

where A is a positive self-adjoint operator and $R(u)$ is a lower order nonlinear nonhomogeneous term.

The dissipative nature of equation (1.4) is reflected in the following properties of the solution map $u_0 \to u(t) = S(t)u_0$:

(i) $\lim_{t\to 0} S(t)u_0 = u_0$ in H, $S(t+s)u_0 = S(t)(S(s)u_0)$, $t, s \geq 0$.
(ii) The map $t \to S(t)u_0$ is analytic from $(0, \infty)$ to H.
(iii) $S(t)$ is injective for $t \geq 0$.
(iv) There exists Y, compact in H, convex and absorbing; i.e., for any bounded set F in H there exists $t_0 = t_0(F)$ such that $S(t)u_0 \in Y$ for all $t \geq t_0$, $u_0 \in F$.

These properties are standard features of autonomous dissipative PDEs and are true for a much larger class of equations than the examples mentioned above. In Chapters 2 to 14, where the term $N(u)$ of (1.5) is not yet specified, properties (i) to (iv) will be assumed to hold.

We denote by $(\Lambda_j)_j$ the increasing sequence of distinct eigenvalues of A and by (λ_j) the nondecreasing sequence of eigenvalues counted with their multiplicities. The linearization around $u(t)$ of $N(u)$ will be denoted by $A(t)$, i.e.,

$$A(t) = A + L(t), \qquad L(t)v = \frac{\partial R}{\partial u}(u(t))v, \tag{1.6}$$

where $\partial R/\partial u$ denotes the Gateaux differential of $R(u(\cdot))$ with respect to $u(\cdot)$.

In Chapter 2 we treat the transport of finite-dimensional contact elements. By a finite-dimensional contact element we mean a pair (u_0, P_0) with $u_0 \in H$ and P_0 a finite-dimensional projector (orthogonal projector operator) in H. One regards P_0 as the projector on the tangent space at u_0 to an infinitesimal surface passing through u_0. The transport under $S(t)$ of this surface induces the transport of (u_0, P_0) according to

$$u(t) = S(t)u_0, \tag{1.7}$$

$$\frac{d}{dt}P(t) + (I - P(t))A(t)P(t) + P(t)A(t)^*(I - P(t)) = 0, \tag{1.8}$$

$$P(0) = P_0, \tag{1.9}$$

where $A(t)$ is the linearized operator given in (1.6) and $A(t)^*$ is its adjoint in H.

In the linear case ($A(t) = A$) one can completely determine the asymptotic behavior of solutions of (1.7) to (1.9) by studying the equation for the trace $\mathrm{Tr}(AP(t))$.

We continue the study of equations (1.7) to (1.9) in Chapter 3. For any n-dimensional contact element (u, P) we introduce the quantities

$$\Lambda(P) = \mathrm{Max}\{(Ag, g) \mid |g| = 1, Pg = g, g \in \mathscr{D}(A)\}, \tag{1.10}$$

$$\lambda(P) = \mathrm{Min}\{(Ag, g) \mid |g| = 1, Pg = g, g \in \mathscr{D}(A)\}, \tag{1.11}$$

where $(\ ,\ )$ and $|\ |$ denote the scalar product and the norm in H; $\mathscr{D}(A)$ is the domain of A. It follows from the minimax and maximin theorems that $\Lambda(P) \geq \lambda_n$, $\lambda(P) \leq \lambda_{n+1}$. These two quantities measure the position of the linear space $\mathrm{Ker}(I - P)$ relative to the fixed orthonormal system of coordinates formed with the eigenvectors $\{w_j\}_{j=1}^{\infty}$ of A ($Aw_j = \lambda_j w_j$). We assume that $L(t)$ satisfy bounds of the type

$$|L(t)v|^2 \leq k_1|v|^2 + k_2|A^{1/4}v|^2 + k_3|A^{1/2}v|^2, \tag{1.12}$$

$$|L(t)^*v|^2 \leq k_1|v|^2 + k_2|A^{1/4}v|^2 + k_3|A^{1/2}v|^2, \tag{1.13}$$

reflecting the fact that $R(u)$ is assumed to be of lower order (half the "number of derivatives" at most) than A. The "constants" k_1, k_2, k_3 (and all other constants denoted by a k in the text) depend through the function $u(t) = S(t)u_0$ on the initial data u_0. Our assumption is that if $u_0 \in \theta Y$, $\theta \geq 1$ (some dilation of Y), then these constants can be chosen uniformly, as functions of θ only. (Of course this involves, in applications, good a priori bounds on $u(t)$ in various Sobolev norms.)

We derive under these assumptions differential inequalities for the transported quantities $\lambda(t) = \lambda(P(t))$, $\Lambda(t) = \Lambda(P(t))$. For instance, if the linear diffusion operator A has gaps in the spectrum that are large with respect to constants k_1, k_2, k_3, more precisely if

$$(\Lambda_{m+1} - \Lambda_m)^2 > k_1 + k_2\left[\frac{\Lambda_m + \Lambda_{m+1}}{2}\right]^{1/2} + k_3\frac{\Lambda_m + \Lambda_{m+1}}{2} \tag{1.14}$$

for some m, then we can deduce the useful

Theorem 1.1 (Spectral Blocking Property; cf. #3.3), *Let* $\lambda(t) = \lambda(P(t))$, $\Lambda(t) = \Lambda(P(t))$ *be defined in* (1.10), (1.11), *for* $P(t)$ *solving* (1.7) *to* (1.9). *Then*

(a) *if for some* $t_0 \geq 0$, $\Lambda(t_0) < (\Lambda_m + \Lambda_{m+1})/2$ *for some* m *satisfying* (1.14), *then* $\Lambda(t) < (\Lambda_m + \Lambda_{m+1})/2$ *for all* $t \geq t_0$,

(b) *if for some* $t_0 \geq 0$, $\lambda(t_0) > (\Lambda_m + \Lambda_{m+1})/2$ *for some (possibly different)* m *satisfying* (1.14), *then* $\lambda(t) > (\Lambda_m + \Lambda_{m+1})/2$ *for all* $t \geq t_0$.

Thus $\lambda(t)$ (resp. $\Lambda(t)$) cannot cross large gaps in the spectrum of A from the right (resp. left).

We note here that although a condition of the type $\lambda(t_0) > (\Lambda_m + \Lambda_{m+1})/2$

can be realized only if the dimension n of $P(t_0)$ is large enough ($\lambda_{n+1} > (\Lambda_m + \Lambda_{m+1})/2$), conditions of the type $\Lambda(t_0) < (\Lambda_m + \Lambda_{m+1})/2$ do not impose restrictions on the dimension of $P(t_0)$ provided the set of m's for which (1.14) is valid is not bounded. In particular, the blocking of $\Lambda(P(t))$ in the $n = 1$ case has important consequences, established in Chapter 4. Let us denote by P_n the spectral projector of A on the span of $w_1, \ldots, w_n$. Let us consider the locally compact cone in H

$$K_n = \left\{ w \in \mathscr{D}(A^{1/2}) \,|\, |A^{1/2}w|^2 \leq \frac{\lambda_n + \lambda_{n+1}}{2} |w|^2 \right\}. \tag{1.15}$$

In Chapter 4 we prove the following strong squeezing properties.

Theorem 1.2 (cf. #4.3). *Let n be large enough. Let $w(t)$ be a solution of the linearized equation around $S(t)u_0 = u(t)$, with $u_0 \in \theta Y$ for some $\theta \geq 1$:*

$$\frac{dw}{dt} + A(t)w = 0, \qquad w(0) = w_0. \tag{1.16}$$

If for some $t_0 \geq 0$, $w(t_0)$ belongs to K_n, then for all $t \geq t_0$, $w(t)$ belongs to K_n. Moreover, the following alternative holds: Either:

(a) $|w(t)| \leq |w(0)| \exp(-kt)$ *for all $t \geq 0$,*

or

(b) *there exists a finite $t_0 > 0$ such that the inequality in* (a) *holds for $t \leq t_0$ and $w(t)$ belongs to K_n for $t \geq t_0$.*

The precise condition on the size of n is given in Theorem 4.2, but essentially the requirement is that $\lambda_n > 5(\Lambda_m + \Lambda_{m+1})$ for some m satisfying the gap conditions (1.14). Using a slight modification of Theorem 1.1, we obtain also

Theorem 1.3 (Strong Squeezing Property; cf. #4.2). *Let n be large enough (same condition as in Theorem 1.2). Let $w(t) = S(t)u_0 - S(t)u$ be the difference of two solutions with u, $u_0 \in \theta Y$ for some $\theta \geq 1$. Then the conclusions of Theorem 1.2 hold for $w(t)$.*

The strong squeezing property was established for the Kuramoto–Sivashinsky equation in [FNST, FNST1].

In Chapter 5 the cone invariance properties are established. For $\gamma > 0$ and $n \geq 1$ we define

$$C_{n,\gamma} = \{ w \in H \,|\, |(I - P_n)w| \leq \gamma |P_n w| \}. \tag{1.17}$$

If a gap condition involving the constant γ but otherwise entirely similar to (1.14) is satisfied, then we obtain

Theorem 1.4 (The Cone Invariance Property; cf. #5.1). *Let $\gamma > 0$, $\theta \geq 1$ be fixed. Let n be large enough. Consider $w(t) = u_1(t) - u_2(t)$ the difference of two*

solutions $u_1(t)$, $u_2(t)$ *of* (1.4) *with initial data* $u_1(0) \in \theta Y$, $u_2(0) \in \theta Y$. *Then, if* $w(0) \in C_{n,\gamma}$ *it follows that* $w(t) \in C_{n,\gamma}$ *for all* $t \geq 0$.

A similar theorem is true for solutions $w(t)$ of the linearized equation (1.16) (see #5.2). The above invariance property of $C_{n,\gamma}$ is called the cone property for equation (1.4)–(1.5). It was considered in [FNST] and then in [FNST1] and [M-PS], where this terminology was introduced. The strong squeezing property, which yields in particular the invariance property in the cone K_n, is somewhat stronger than the cone invariance property. Indeed, on one hand K_N is locally compact in H, and on the other hand $K_N \subset C_{n,\gamma}$ if

$$\frac{\lambda_N + \lambda_{N+1}}{2} \leq \frac{\gamma^2}{1+\gamma^2} \lambda_{n+1}. \tag{1.17a}$$

The consequences of this property regarding the universal attractor are studied in Chapter 6. We prove

Theorem 1.5 (cf. #6.1). *If* n *is large enough, then the projector* P_n *is injective when restricted to the universal attractor* X *and its inverse is Lipschitz. More precisely,*

$$|(I - P_n)(x - y)| \leq \tfrac{1}{3}|P_n(x - y)|$$

for every x, y *in* X.

It is known [M] that because X has finite fractal dimension, most projectors are injective on X; however, P_n is an important explicit one with this property. Therefore, Theorem 1.5 is interesting although it easily follows from Theorem 1.3. The requirements on n are essentially that (1.14) be satisfied for a gap at Λ_m where $\lambda_N = \Lambda_m$ and λ_N satisfies (1.17) for $\gamma = \frac{1}{3}$.

Denoting

$$C_{n,X} = \bigcap_{x \in X} \{u \in H \,|\, |(I - P_n)(u - x)| \leq \tfrac{1}{3}|P_n(u - x)|\},$$

we deduce from Theorem 1.3 that $S(t)C_{n,X} \subset C_{n,X}$ if n is large enough, that $X \subset C_{n,X}$ (Theorem 1.5), and that as long as a solution $S(t)u_0$ remains in the complement of $C_{n,X}$ its distance to X decreases exponentially. Finally, we conclude Chapter 6 by showing that the complement of a large ball in P_nH is included in $C_{n,X}$.

In Chapter 7 we consider a smooth n-dimensional positively invariant surface Σ. We assume that it is "blocked" in the sense that $\lambda(u) > (\lambda_m + \lambda_{m+1})/2$ for $u \in \Sigma$ and $\lambda(u) = \lambda(P(u))$ with $P(u)$ the projector on the tangent space at u to Σ; here as well as in the sequel it is assumed that n is the dimension of $P(u)$ and that $\lambda_n < \lambda_{n+1}$. We show that under these assumptions, as long as the distance from some solution $S(t)u_0$ to Σ is attained on Σ, it must decay exponentially (at an explicit uniform rate). Also we recall, in Chapter 8, some results concerning the exponential decay of volume elements and estimates on the Hausdorff and the fractal dimensions of X (see [CF, CFMT, CFT]).

Chapter 9 is devoted to the description of the initial data for our construction of inertial manifolds. They form the smooth oriented boundary Γ of a bounded, open, connected set D included in $P_n H$; n is chosen sufficiently large. We denote at each $u \in \Gamma$ by $P(u)$ the projector on the space $\mathbb{R}N(u) + T_u(\Gamma)$, where $T_u(\Gamma)$ is the tangent space at u to Γ, and by $\nu(u)$ the outward unit normal in $P_n H$ to Γ, and we set $\lambda(u) = \lambda(P(u))$, $\Lambda(u) = \Lambda(P(u))$. Then the properties of Γ are

(I) $\Lambda(u) < (\lambda_n + \lambda_{n+1})/2$ for any $u \in \Gamma$.
(II) $\lambda(u) > (\lambda_n + \lambda_{n+1})/2$ for any $u \in \Gamma$.
(III) $(N(u), \nu(u)) > 0$ for any $u \in \Gamma$.
(IV) $\Gamma \subset C_{n,X}$.
(V) For any $u \in \Gamma$, $\mathbb{R}N(u) + T_u(\Gamma) \subset C_{n,\gamma}$ for some $\gamma > 0$.

Properties (I) and (II) assert that the initial surface Γ is "spectrally blocked." Property (III) shows that $d(S(t)u_0)/dt|_{t=0+}$ at any $u_0 \in \Gamma$ points toward the interior of D. In the applications, Γ is usually a simple explicit set: a large sphere for the Kuramoto–Sivashinsky equation, a large ellipsoid for the Burgers and reaction–diffusion equations, a large level set of the Lyapunov function in the case of the Cahn–Hilliard equation. The construction of Γ is a purely geometric problem. The feasibility of (III) depends on a coercivity property of $N(u)$ for large u. Conditions (I) and (II) can be achieved provided the gap $\lambda_{n+1} - \lambda_n$ is large enough in comparison to the size of the nonlinear term of Γ (see 9.10), (9.12)).

In Chapter 10 we use the spectral blocking, strong squeezing, and volume decay properties in order to construct the inertial manifold starting from Γ. We denote by Σ the integral manifold having Γ as initial data:

$$\Sigma = \bigcup_{t>0} S(t)\Gamma. \tag{1.18}$$

We establish first, using (I) and the spectral blocking property, the fact that the projection P_n on Σ is a regular map (has invertible Jacobian) at any point of Σ. From the results in Chapter 5 and condition (IV) for Γ it follows that $\Sigma \subset C_{n,X}$; thus $P_n\Sigma \cap P_n X$ is empty since we take Γ far away from X. We show that the closure $\overline{P_n\Sigma}$ of $P_n\Sigma$ is included in the union of the disjoint sets $P_n X$, $P_n\Sigma$, and Γ. We use next the isoperimetric inequality and the exponential decay of surfaces of dimension larger than or equal to $n - 1$ to show that $\overline{P_n\Sigma} \supset \bar{D}$. Since P_n is regular on Σ and since Σ is connected, we deduce that P_n restricted to Σ is injective. It follows that $\bar{D} = P_n\Sigma \cup P_n X \cup \Gamma$ and we can define on $\bar{D}$ the inverse Φ of the restriction of P_n to $\bar{\Sigma}$, $\Phi : \bar{D} \to \bar{\Sigma}$, $\Phi|_\Gamma =$ identity. We show, using the strong squeezing property and (V) (see Theorem 1.1), that

$$|(I - P_n)(\Phi(p_1) - \Phi(p_2)| \leq \tfrac{1}{3}|P_n(\Phi(p_1) - \Phi(p_2)| \tag{1.19}$$

for any p_1, p_2 in D.

Finally, we show using (II), the spectral blocking, and results in Chapters 5 and 7 that for any u_0, $\operatorname{dist}(S(t)u_0, \bar{\Sigma})$ decreases exponentially. We conclude that $\bar{\Sigma}$ is an inertial manifold also satisfying, besides properties (1.1), (1.2), (1.3),

$\bar{\Sigma}$ is the graph of an explicit Lipschitz map. (1.20)
$\bar{\Sigma}$ is the closure of a smooth manifold. (1.21)
The n-dimensional volume of $\bar{\Sigma}$ is finite. (1.22)
Relations (I) and (II) hold for any $u \in \Sigma$, where $P(u)$ denotes the orthogonal projector on the tangent space at u to Σ. (1.23)

In Chapter 11 we establish a consequence of the existence of the inertial manifolds constructed previously: we give a lower bound on the exponential rate of convergence to the attractor. Let B be a large ball in H. Let us denote $\delta_X(t) = \sup\{\text{dist}(u, X) | u \in S(t)B\}$. We prove that if n is the dimension of an inertial manifold constructed by our method, then

$$\delta_X(t) \geq \exp\left[-\frac{C_n t}{n - d_M(X)}\right] \tag{1.24}$$

for all t large, where C_n is an explicit constant and $d_M(X)$ is the fractal dimension of X. (In general $C_n/(n - d_M(X))$ is smallest when n is as small as possible.) Since we can construct inertial manifolds of arbitrarily large dimensions, condition (1.24) is not vacuous. The proof of (1.24) uses in a fundamental way the information that $\bar{\Sigma}$ is the closure of a smooth finite-dimensional Euclidean manifold.

Once we have proved the existence of inertial manifolds it is clear that this new tool will be useful if it possesses a certain number of interesting properties, in particular properties that are not satisfied by the universal attractor which otherwise probably describes the dynamics conveniently. Two such properties are already proved at this point: the Lipschitz regularity of the inertial manifold (to be compared to the fractal nature of the attractor) and, on the other hand, the asymptotic convergence of the trajectories to the manifold at an exponential rate (while the convergence of the trajectories to the universal attractor occurs at an unspecified rate). Besides these properties, we prove in Chapters 12, 13, and 14 two important properties that are satisfied by the inertial manifolds and are not usually satisfied by the attractors. The first one, proved in Chapters 12 and 13, is *asymptotic completeness*, and the second one, proved in Chapter 14, is the *stability* with respect to perturbations. The asymptotic completeness of $\bar{\Sigma}$ means that given any orbit $u(t)$, we can find $t_0 > 0$ and $u_0 \in \bar{\Sigma}$ such that $u(t + t_0) - S(t)u_0$ converges to 0 as $t \to \infty$; since the orbit $S(t)u_0$ is included in $\bar{\Sigma}$ this means that given any orbit $u(t)$ of the dynamical system, there exists an orbit ($= S(t)u_0$) lying in $\bar{\Sigma}$ that produces the same asymptotic behavior for $t \to \infty$ (see #12.1).

The proof of this result is rather laborious and takes two chapters. In Chapter 12 we prove that the map Φ parametrizing the inertial manifold $\bar{\Sigma}$ is Lipschitz in a stronger norm:

$$|(I - P_n)A^{1/2}(\Phi(p_1) - \Phi(p_1))| \leq \frac{1}{3}\left[\frac{\lambda_n + \lambda_{n+1}}{2}\right]^{1/2} |p_1 - p_2| \quad \text{cf. \#12.1).} \tag{1.25}$$

Next we use (1.25) to give an estimate of the maximal rate at which trajectories lying on $\bar{\Sigma}$ can separate backward in time:

$$|u_1(t) - u_2(t)| \le \frac{\sqrt{10}}{3} |P_n(u_1(0) - u_2(0))| e^{\mu|t|}, \qquad t < 0, \tag{1.26}$$

$$\mu = \lambda_n + \text{lower order terms.} \quad (\text{See } \#12.3) \tag{1.27}$$

The last result (i.e., #12.4) of Chapter 12 is a bound on the minimal rate of convergence to $\bar{\Sigma}$ of solutions lying in $C'_{n,X} = \{u \in H | u - x \in C_{n,1/2}$ for all $x \in X\}$:

$$\operatorname{dist}(u(t), \bar{\Sigma}) \le \operatorname{dist}(u(0), \bar{\Sigma}) e^{-\sigma t}, \qquad t \ge 0, \tag{1.28}$$

$$\sigma = \frac{\lambda_{n+1} + \lambda_n}{2} - \text{lower order terms.} \tag{1.29}$$

Comparing (1.27) and (1.29) one remarks that

$$\sigma - \mu = \frac{\lambda_{n+1} - \lambda_n}{2} + \text{lower order terms.}$$

This means that the convergence (forward in time) of solutions (even in the "slow" cones $C'_{n,X}$) toward $\bar{\Sigma}$ occurs at a much higher exponential rate than the separation (backward in time) of trajectories on $\bar{\Sigma}$. This is the main reason for the success of the argument finishing the proof of asymptotic completeness. This argument, given in Chapter 13, is the adaptation of the following heuristic algorithm, familiar in the theory of local center manifolds for ODEs (see [Ha, Ca]): Let $u(t)$ be any solution of (1.4). Wait until $P_n u(t)$ belongs to $P_n(\Sigma)$ (this happens if t is large). Find $\tilde{u} \in \Sigma$ such that $P_n u(t) = P_n \tilde{u}$. Run back in time on $\bar{\Sigma}$ starting from $\tilde{u}$ until you hit Γ, at $v = v(t)$. Increase t and repeat the procedure. Then $v(t)$ will converge to some $v_0 \in \Gamma$. The trajectory on $\bar{\Sigma}$ emanating from v_0 will be the sought companion trajectory to $u(t)$.

The stability property proved in Chapter 14 is a kind of continuity property of $\bar{\Sigma}$ with respect to perturbations: we consider suitable perturbations of the viscosity parameter ν and of the driving force f; we also consider perturbations of the operators corresponding to their finite-dimensional Galerkin approximations. In each case we show that the perturbed system also possesses an inertial manifold $\bar{\Sigma}_m$ and that, as the parameter $m \to \infty$, $\bar{\Sigma}_m$ converges to $\bar{\Sigma}$ in an appropriate sense.

Chapters 15 to 19 are devoted to explicit examples. In this work we do not insist on the best estimates or the most general examples. Rather, we wish to illustrate the range of applicability of our method.

In Chapter 15 we study the Kuramoto–Sivashinsky equation

$$\frac{\partial v}{\partial t} + \frac{\partial^4 v}{\partial x^4} + \frac{\partial^2 v}{\partial x^2} + v\frac{\partial v}{\partial x} = 0 \tag{1.30}$$

for an odd periodic function v on $[-L/2, L/2]$, $L > 0$. In [FNST, FNST1] it was proved by a fixed-point argument in a functional space that (1.30) possesses a Lipschitz inertial manifold. We use the change of variable $u + \varphi = v$ with the explicit time-independent φ used in [FNST, FNST1]. We prove that a sphere $\Gamma = \{u | P_n u = u, \ |u| = R\}$ with R large, n of the order of L^7, satisfies conditions (I) to (V) in the L^2 space of odd periodic functions. The dimension of the inertial manifold that we construct is thus of the order L^7. This is a weaker result from this point of view than [FNST, FNST1]. However, our present construction is more explicit and is based on a transparent connection with the forward (well-posed) initial value problem for (1.30). Also, our proof provides more insight into the geometric nature of the inertial manifolds and the dynamics of the Kuramoto–Sivashinsky equation. These dynamics are surprisingly complicated, as indicated by the recent numerical studies [HN] and [HNZ], even for low values of L. We hope that our present inertial manifold $\bar{\Sigma}$ (due to its very explicit construction) may subsist even below the high dimensions requested by our theorem.

In Chapter 16 we apply the methods developed so far to the nonlocal Burgers equation

$$\frac{\partial u}{\partial t} - \frac{\partial^2 u}{\partial x^2} + \frac{\partial u}{\partial x} \int_0^L \omega(y) u(y)\, dy = f \tag{1.31}$$

where u is L-periodic, $\int_0^L u\, dx = 0$, and f, $\omega \in L^2((0, L))$. This equation is interesting, because unlike the classical local Burgers equation, it can display for certain adequate choices of ω and f nontrivial dynamics; an example is provided at the end of the chapter. It turns out that in this case the role played by condition (1.14) is replaced by the much weaker condition

$$\left[\frac{\Lambda_{m+1} - \Lambda_m}{2}\right]^2 > k_1 \left(= \frac{L^2}{2\pi^2} \left[\int_0^L \omega^2\, dy\right] \left[\int_0^L f^2\, dy\right]\right). \tag{1.32}$$

Assuming also $f'' \in L^2((0, L))$, we obtain that (1.31) has an n-dimensional inertial manifold provided

$$n > c \cdot \max\left\{\left[\int_0^L \omega^2\, dy\right]^{1/2} \left[\int_0^L f^2\, dy\right]^{1/2} L^3, L^2 \left[\frac{\int_0^L f''^2\, dy}{\int_0^L f^2\, dy}\right]^{1/2}\right\} \tag{1.32a}$$

with an adequate absolute constant c.

In the second part of Chapter 16 we show that if ω belongs to a certain dense set of H, then our construction can be implemented for a dimension independent of the driving force f, which can be much smaller than that given by the estimate (1.32a). The resulting manifold may not be smooth but is inertial (i.e., has all the other properties of the inertial manifolds considered so far) (see #16.6).

In Chapter 17 we apply again in a slightly modified form the methods developed in Chapters 2 to 12 to the following Cahn–Hilliard equation:

$$\begin{gathered} \frac{\partial u}{\partial t} + \frac{\partial^2 u}{\partial x^4} + \frac{\partial^2}{\partial x^2} p(u) = p, \quad \text{with} \\ p(u) = -b_2 L^{-2} u - b_3 L^{-1} u^2 - b_4 u^3, \end{gathered} \tag{1.33}$$

where u is L-periodic, $\int_0^L u\,dx \equiv 0$, b_2, b_3, b_4 are constants, $b_2 < 0 < b_4$. This equation possesses a Lyapunov functional, namely

$$V(u) = \frac{1}{2}\int_0^L u'^2\,dy + \frac{b_2}{2L^2}\int_0^L u^2\,dy + \frac{b_3}{3L}\int_0^L u^3\,dy + \frac{b_4}{4}\int_0^L u^4\,dy. \tag{1.34}$$

Thus the universal attractor is formed by the stationary solutions together with the heteroclinic and homoclinic trajectories (see also [BV]). However, this set may still be very complicated. Therefore the search for inertial manifolds is meaningful. In this case again a condition similar to (1.32) plays the role of condition (1.14). There is still another modification that must be made here. Indeed, in this case the natural choice of Γ is

$$\Gamma = \{u = P_n u : V(u) = bL^{-3}\}$$

with an appropriate large constant b and dimension $n - 1$. With this choice, the closure $\bar{\Sigma}$ of the integral manifold defined in (1.18) is an inertial manifold for the Cahn–Hilliard equation (1.33).

In Chapter 18 we apply our method to equations of the type

$$\frac{\partial u}{\partial t} - \Delta u = f(u) + g \tag{1.35}$$

in $[-\pi, \pi]^2$ with periodic boundary conditions. The function g is smooth and odd, f is odd, smooth, $f'(u) \le \delta < 1$, and

$$|f'(u)| \le c(1 + |u|^p), \qquad |f''(u)| \le c(1 + |u|^p), \qquad 0 < p < 1.$$

We quickly review a proof of global existence. The spectral blocking property holds because in this case $k_2 = k_3 = 0$ and one can find arbitrarily large gaps between consecutive sums of two squares. The initial data set Γ is chosen a large ellipsoid $\Gamma = \{u | P_n u = u, |A^{1/2}u| = R\}$ with n and R large enough and

$$|A^{1/2}u|^2 = \int_{-\pi}^{\pi}\int_{-\pi}^{\pi} |\nabla u|^2\,dx\,dy.$$

The condition $f'(u) \le \delta < 1$ forces the dynamics to be trivial: there is a unique globally attracting stationary solution. However, our method applies with no difficulty to the case when $f'(u) \le \delta < 1$ is required to hold for large u and also to the case in which u and $f(u)$ are vector valued. In these cases the universal attractor can be complicated. (See [CHS] for an interesting pioneering example.)

In Chapter 19, as a specific example of such relaxed conditions on the nonlinearity, we treat the Chaffee–Infante reaction–diffusion equation in $[-\pi, +\pi]^2$ with periodic boundary conditions and without restricting ourselves to odd solutions:

$$\frac{\partial u}{\partial t} - \Delta u + \lambda(u^3 - u) = 0 \tag{1.36}$$

where $\lambda > 0$ is the bifurcation parameter. Generically in λ, the universal attractor consists of multiple steady states and their unstable manifolds. The weak maximum principle and nonlinear Gronwall inequalities are used to obtain absorbing sets; the absorption time is shown to be independent of initial data. The noncoercivity of A is easily overcome by the choice of the initial data set Γ, which is a large ellipsoid

$$\Gamma = \{u | P_n u = u, |A^{1/2}u|^2 + \lambda |u|^2 = R^2\}$$

with n and R large enough and $|A^{1/2}u|^2$ as in Chapter 18. This chapter can easily be extended to cover analogs of (1.36) with more general cubic-like nonlinearities.

The integral manifold method for the construction of inertial manifolds developed in this work is well suited for the numerical computations of inertial manifolds. Although we do not study this aspect in the present work, we make a few remarks. The main requirement for the numerical computation of an inertial manifold by our method is the existence of fast and robust code integrating evolution equations, with sufficient relative tolerance. Our method avoids any backward-time integration instabilities and shortcuts any fixed-point algorithms. The construction of fast and robust algorithms now hinges upon a careful choice of (i) the initial surface Γ, (ii) the discrete grid on Γ initiating a discrete set of integral curves on Σ, and (iii) completing the interpolation of Σ. These considerations will be developed in a further work.

CHAPTER 2

The Transport of Finite-Dimensional Contact Elements

Let $u(t) = S(t)u_0$ be a solution of

$$\frac{du}{dt} + N(u) = \frac{du}{dt} + Au + R(u) = 0, \tag{2.1}$$

$$u(0) = u_0, \tag{2.2}$$

where we choose $R(u)$ of the form $R(u) = B(u, u) + Cu + f$, with a constant $f \in H$, a linear operator C, and a bilinear operator B (both of lower differential order than A).

Although this is not the general case to which our considerations apply, it is an adequate sample that is simple enough and already displays all the interesting nonlinear features.

By a finite-dimensional contact element we mean a pair (u, P) with $u \in H$ and P an n-dimensional orthogonal projector in H. The space $u + \text{Ker}(I - P)$ which can be viewed as the tangent hyperplane at $S(t)u_0$ to $S(t)\Sigma$, will be spanned by the vectors $v_1(t), \ldots, v_n(t)$, solutions of the equations

$$\frac{dv_i}{dt} + A(t)v_i = 0, \tag{2.3}$$

$$v_i(0) = v_i^0, \qquad i = 1, 2, \ldots, n, \tag{2.4}$$

with

$$A(t)w = Aw + B(u(t), w) + B(w, u(t)) + Cw. \tag{2.5}$$

For later use let us denote the lower order terms in $A(t)$ by $L(t)$:

$$L(t)w = B(u(t), w) + B(w, u(t)) + Cw.$$

Let $\hat{v}(t)$ be defined by

$$\hat{v}(t) = v_1(t) \wedge v_2(t) \wedge \cdots \wedge v_n(t); \tag{2.6}$$

$\hat{v}(t)$ is an element of the Hilbert space $\Lambda^n H$ endowed with the scalar product, defined on the generating set $\{\omega_1 \wedge \omega_2 \wedge \cdots \wedge \omega_n | \omega_1, \ldots, \omega_n \in H\}$ by

$$(\hat{v}, \hat{\omega}) = \det[(v_j, \omega_k)]_{j,k=1}^n$$

(where $(\cdot,\cdot)$ denotes the usual scalar product in H) and norm $|\hat{v}| = (\hat{v}; \hat{v})$. We shall recall a few facts concerning $\hat{v}(t)$ (see [CF] for details). From (2.3), (2.4) it follows that $\hat{v}(t)$ evolves according to the equation

$$\frac{d\hat{v}(t)}{dt} + A_n(t)\hat{v}(t) = 0, \tag{2.7}$$

$$\hat{v}(0) = \hat{v}_0 = v_1^0 \wedge \cdots \wedge v_n^0, \tag{2.8}$$

where for an operator $T: \mathscr{D}(T) \subset H \to H$, we defined the operator T_n in $\Lambda^n H$ by

$$T_n = T \wedge I \wedge \cdots \wedge I + I \wedge T \wedge \cdots \wedge I + \cdots I \wedge \cdots \wedge I \wedge T.$$

If $\hat{\omega} = \omega_1 \wedge \cdots \wedge \omega_n$ in $\Lambda^n H$ and $\omega_i \in \mathscr{D}(T)$, $i = 1, 2, \ldots, n$, then

$$(T_n\hat{\omega}; \hat{\omega}) = \operatorname{Tr}(TP_\omega)|\hat{\omega}|^2 \quad \text{and} \tag{2.9}$$

$$(T_n\hat{\omega}; T_n\hat{\omega}) = |\hat{\omega}|^2 | \operatorname{Tr} T^*(I - P_\omega)TP_\omega + (\operatorname{Tr} TP_\omega)^2] \tag{2.10}$$

where P_ω is the projector on the span of $\omega_1, \ldots, \omega_n$. One can check (2.9) and (2.10) by reducing the proof to the case when $\{\omega_i\}$ is an orthonormal system formed by eigenvectors of either $P_\omega T$ or $P_\omega T^*(I - P_\omega)T$ restricted to $P_\omega H$, and by splitting the operator $T = (I - P_\omega)T + P_\omega T$.

Let us denote by $P(t)$ the projector on the space spanned by $v_1(t), \ldots, v_n(t)$.

Theorem 2.1. *The evolution equation satisfied by the contact element* $(u(t), P(t))$ *is*

$$\frac{d}{dt}P(t) + (I - P(t))A(t)P(t) + P(t)A(t)^*(I - P(t)) = 0, \tag{2.11}$$

$$P(0) = P_0 = \textit{projector on the span of } v_1^0, \ldots, v_n^0, \textit{ with } A(t) \textit{ given in (2.5)}, \tag{2.12}$$

$$u(t) = S(t)u_0.$$

Proof. Let v_0 be a fixed element in H. Then

$$v_0 \wedge \hat{v}(t) = (I - P(t))v_0 \wedge \hat{v}(t)$$

and

$$|v_0 \wedge \hat{v}(t)|^2 = |(I - P(t))v_0|^2 |\hat{v}(t)|^2.$$

Now, from (2.7) and (2.9) it follows that

$$\frac{1}{2}\frac{d}{dt}|\hat{v}(t)|^2 = -\mathrm{Tr}(A(t)P(t))|\hat{v}(t)|^2. \tag{2.13}$$

On the other hand, $v_0 \wedge \hat{v}(t)$ satisfies the equation

$$\frac{d}{dt}(v_0 \wedge \hat{v}(t)) + A_{n+1}(t)(v_0 \wedge \hat{v}(t)) - A(t)v_0 \wedge \hat{v}(t) = 0.$$

Therefore

$$\frac{1}{2}\frac{d}{dt}|v_0 \wedge \hat{v}(t)|^2 = -\mathrm{Tr}(A(t)Q(t))|v_0 \wedge \hat{v}(t)|^2 + (A(t)v_0 \wedge \hat{v}(t); v_0 \wedge \hat{v}(t)) \tag{2.14}$$

where $Q(t)$ is the projector on the span of $v_0, v_1(t), \ldots, v_n(t)$. We remark that the term $(A(t)v_0 \wedge \hat{v}(t); v_0 \wedge \hat{v}(t))$ is equal to $((I - P(t))A(t)v_0, v_0)|\hat{v}(t)|^2$. Since

$$|(I - P(t))v_0|^2 = \frac{|v_0 \wedge \hat{v}(t)|^2}{|\hat{v}(t)|^2}$$

it follows from (2.13), (2.14) that

$$\begin{aligned}
\frac{1}{2}\frac{d}{dt}|(I - P(t))v_0|^2 &= -\mathrm{Tr}(A(t)Q(t))|(I - P(t))v_0|^2 \\
&\quad + ((I - P(t))A(t)v_0, v_0) + \mathrm{Tr}(A(t)P(t))|(I - P(t))v_0|^2 \\
&= -\mathrm{Tr}(A(t)(Q(t) - P(t)))|(I - P(t))v_0|^2 + ((I - P(t))A(t)v_0, v_0) \\
&= -(A(t)(I - P(t))v_0, (I - P(t))v_0) + ((I - P(t))A(t)v_0, v_0) \\
&= (A(t)P(t)v_0, (I - P(t))v_0) = ((I - P(t))A(t)P(t)v_0, v_0).
\end{aligned}$$

From the parallelogram identity it follows at once that

$$\frac{d}{dt}(I - P(t)) = (I - P(t))A(t)P(t) + P(t)A(t)^*(I - P(t))$$

and thus the evolution equation (2.11) is established. □

One of the important quantities associated to $P(t)$ is $\mathrm{Tr}(AP(t))$. In order to derive the equation satisfied by $\mathrm{Tr}(AP(t))$, let us note that (2.9) gives the identity

$$\mathrm{Tr}(AP(t)) = \frac{(A_n\hat{v}(t); \hat{v}(t))}{|\hat{v}(t)|^2}. \tag{2.15}$$

Now, denoting $L_n(t) = (L(t))_n$,

$$\frac{1}{2}\frac{d}{dt}(A_n\hat{v}(t); \hat{v}(t)) = -(A_n\hat{v}(t); A_n\hat{v}(t)) - (A_n\hat{v}(t); L_n(t)\hat{v}(t)). \tag{2.16}$$

Combining (2.16) with (2.13), we obtain from (2.15)

$$\frac{1}{2}\frac{d}{dt}\operatorname{Tr}(AP(t)) = -\frac{(A_n\hat{v}(t); A_n\hat{v}(t))}{|\hat{v}(t)|^2} + \frac{(A_n\hat{v}(t), \hat{v}(t))}{|\hat{v}(t)|^2}(\operatorname{Tr} AP(t) + \operatorname{Tr} L(t)P(t))$$
$$- \frac{(A_n\hat{v}(t); L_n(t)\hat{v}(t))}{|\hat{v}(t)|^2}.$$

This last expression can be written as

$$\begin{aligned}\frac{1}{2}\frac{d}{dt}\operatorname{Tr} AP(t) = &-\left|(A_n - \operatorname{Tr} AP(t))\frac{\hat{v}(t)}{|\hat{v}(t)|}\right|^2 \\ &- \left[(A_n - \operatorname{Tr} AP(t))\frac{\hat{v}(t)}{|\hat{v}(t)|}; L_n(t)\frac{\hat{v}(t)}{|\hat{v}(t)|}\right].\end{aligned} \tag{2.17}$$

An immediate consequence of (2.17), noted here for later convenience, is

$$\begin{aligned}\frac{d}{dt}\operatorname{Tr} AP(t) \le &-\left|(A_n - \operatorname{Tr} AP(t))\frac{\hat{v}(t)}{|\hat{v}(t)|}\right|^2 \\ &+ \left[L_n(t)\frac{\hat{v}(t)}{|\hat{v}(t)|}; L_n(t)\frac{\hat{v}(t)}{|\hat{v}(t)|}\right].\end{aligned} \tag{2.18}$$

In order to illustrate the significance of $\operatorname{Tr} AP(t)$ and of (2.17) we conclude this chapter by showing how one can completely determine the asymptotic behavior of contact elements in the case of linear equations. The treatment follows ideas from [FS, FS1]; on the other hand, the following considerations played an inspiring role in [FS2].

We assume thus that $L(t) = 0$. Let us denote $\operatorname{Tr} AP(t) = \mu(t)$, $\hat{v}(t)/|\hat{v}(t)| = \hat{\omega}(t)$. Equations (2.7) and (2.13) on the one hand and (2.17) on the other hand become, respectively,

$$\frac{d}{dt}\hat{\omega}(t) + (A_n - \mu(t))\hat{\omega}(t) = 0 \quad \text{in } \Lambda^n H, \tag{2.19}$$

and

$$\frac{1}{2}\frac{d}{dt}\mu(t) + |(A_n - \mu(t))\hat{\omega}(t)|^2 = 0. \tag{2.20}$$

We recall that $\mu(t) = (A_n\hat{\omega}(t); \hat{\omega}(t))$ and $|\hat{\omega}(t)| = 1$. Let $\hat{\omega}(0) = \hat{\omega}_0$ be the initial data for (2.19). Also let $(m_k)_{k=1,2,\ldots}$ be the sequence of the distinct eigenvalues of the operator A_n. (They are completely determined by the eigenvalues $(\lambda_n)_{j\in N}$ of A by $m_k = \lambda_{i_1} + \cdots + \lambda_{i_n}$ for some $0 < i_1 < i_2 < \cdots < i_n$.)

The first fact we infer from (2.20) is that $\mu(t)$ is nonincreasing and therefore there exists m

$$m_1 \le m = \lim_{t\to\infty} \mu(t) = \inf_{t\to\infty} \mu(t).$$

Integrating in (2.20) we obtain that

$$\frac{1}{2}(m - \mu(0)) + \int_0^\infty |(A_n - \mu(t))\hat{\omega}(t)|^2 \, dt = 0$$

and in particular, since $|(A_n - \mu(t))\hat{\omega}(t)|^2$ is integrable, there exists a sequence $t_j \to \infty$ such that

$$(A_n - \mu(t_j))\hat{\omega}(t_j) \to 0 \quad \text{in } \Lambda^n H. \tag{2.21}$$

Now $|\hat{\omega}(t_j)| = 1$ and therefore we may extract a subsequence of t_j (denoted again t_j) on which $\hat{\omega}(t_j)$ converges weakly to an element $\hat{\omega}$ of $\Lambda^n H$. Since $A_n^{1/2}$ is compact and since $\mu(t_j) = |A_n^{1/2}\hat{\omega}(t_j)|^2$ is bounded, we may assume that the convergence of $\hat{\omega}(t_j)$ to $\hat{\omega}$ is strong in $\Lambda^n H$. Passing to the limit in (2.21) we infer that

$$(A_n - m)\hat{\omega} = 0. \tag{2.22}$$

But $|\hat{\omega}| = 1$, so it follows that m is an eigenvalue of A_n, say $m = m_k$, $1 \le k < \infty$. Let us consider now the spectral projectors in $\Lambda^n H$ R_-, R_k, R_+ corresponding to the sets $[m_1, m_{k-1}]$, $\{m_k\}$, $[m_{k+1}, \infty)$, respectively. (If $k = 1$ we take $R_- = 0$.) For each R that commutes with A_n we obtain from (2.19) that

$$\frac{d}{dt} R\hat{\omega}(t) + (A_n - \mu(t))R\hat{\omega}(t) = 0 \tag{2.23}$$

and

$$\frac{1}{2}\frac{d}{dt}|R\hat{\omega}(t)|^2 + ((A_n - \mu(t))R\hat{\omega}(t); R\hat{\omega}(t)) = 0, \qquad t \ge 0. \tag{2.24}$$

In the case $R = R_-$ we observe that, since $m_k \le \mu(t)$ for all $t \ge 0$,

$$((A_n - \mu(t))R_-\hat{\omega}(t); R_-\hat{\omega}(t)) \le -(m_k - m_{k-1})|R_-\hat{\omega}(t)|^2.$$

Integrating (2.24) between $0 \le t < t_2$ we obtain

$$|R_-\omega(t_2)|^2 \ge |R_-\omega(t)|^2 \exp 2(m_k - m_{k-1})(t_2 - t).$$

But $|R_-\omega(t_2)| \le |\omega(t_2)| = 1$ and thus, letting $t_2 \to \infty$, we infer that necessarily

$$R_-\hat{\omega}(t) \equiv 0 \quad \text{for all } t \ge 0.$$

This implies that the value of m_k is determined by the initial data: It is the smallest eigenvalue corresponding to a nonzero term in the eigenvector expansion of $\hat{\omega}_0$. (If R_l are the spectral projectors on the eigenspaces corresponding to m_l, the value of m_k is determined by

$$k = \min\{l \,|\, |R_l\hat{\omega}_0| \ne 0\}.$$

We now apply (2.24) to $R = R_+$. Since $\lim_{t\to\infty} \mu(t) = \lim_{t\to\infty} \mu(t) = m_k$ there exists $t_0 \ge 0$ such that, for $t \ge t_0$,

$$m_k \le \mu(t) \le m_k + \tfrac{1}{3}(m_{k+1} - m_k) < m_{k+1}.$$

Integrating (2.24) between t_0 and $t \geq t_0$ we get

$$|R_+\hat{\omega}(t)|^2 \leq |R_+\hat{\omega}(t_0)|^2 \exp(-\tfrac{2}{3}(m_{k+1} - m_k)(t - t_0)).$$

It follows that $R_+\hat{\omega}(t)$ decays exponentially. Finally, in the case $R = R_k$, equation (2.23) becomes

$$\frac{d}{dt}(R_k\hat{\omega}(t)) + (m_k - \mu(t))R_k\hat{\omega}(t) = 0$$

and (2.25)

$$R_k\hat{\omega}(t) = e^{\int_0^t (\mu(z)-m_k)\,dz} R_k\hat{\omega}(0).$$

Since $|R_k\hat{\omega}(t)| \leq |\hat{\omega}(t)| = 1$ it follows that

$$0 \leq \int_0^t (\mu(z) - m_k)\,dz \leq \log\frac{1}{|R_k\hat{\omega}(0)|}.$$

Thus the limit of $R_k\hat{\omega}(t)$ exists and is $e^{\int_0^\infty (\mu(z)-m_k)\,dz} R_k\hat{\omega}_0$. We proved, therefore,

Proposition 2.2. *With the preceding notation we have the following relations*:

$$\lim_{t\to\infty} \hat{\omega}(t) = e^{\int_0^\infty (\mu(z)-m_k)\,dz} R_k\hat{\omega}_0 = \frac{R_k\hat{\omega}_0}{|R_k\hat{\omega}_0|}, \tag{2.26}$$

$$\lim_{t\to\infty} (A_n\hat{\omega}(t), \hat{\omega}(t)) = m_k, \quad \textit{with } k = \min\{l \mid R_l\hat{\omega}_0 \neq 0\}, \tag{2.27}$$

$$\begin{gathered} \hat{\omega}(t) = e^{\int_0^t (\mu(z)-m_k)\,dz} R_k\hat{\omega}_0 + R_+\hat{\omega}(t) \quad \textit{with} \\ |R_+\hat{\omega}(t)| \exp(-(m_{k+1} - m_k)t/2) \to 0 \quad (\text{for } t \to \infty). \end{gathered} \tag{2.28}$$

CHAPTER 3

Spectral Blocking Property

Our aim is to investigate the time evolution of the position of $\operatorname{Ker}(I - P(t))$ relative to the fixed system of coordinates given by $w_1, w_2, \ldots, w_n, \ldots$ (the eigenvectors of A).

Definition 3.1. Let P be an n-dimensional projector. We define $\Lambda(P)$, $\lambda(P)$ by

(a) $\Lambda(P) = \max\{(Ag, g) \mid |g| = 1, g \in \mathscr{D}(A), Pg = g\}$,
(b) $\lambda(P) = \min\{(Ag, g) \mid |g| = 1, g \in \mathscr{D}(A), Pg = 0\}$.

It follows from the minimax and maximin theorems [DS] that

$$\lambda(P) \leq \lambda_{n+1}, \qquad \Lambda(P) \geq \lambda_n. \tag{3.1}$$

In order to simplify notation we shall write $\Lambda(t)$ (resp. $\Lambda(t)$) instead of $\Lambda(P(t))$ (resp. $\lambda(P(t))$).

We shall now consider the time evolution of $\Lambda(t)$, $\lambda(t)$ as $P(t)$ evolves according to (2.11). First let us study $\Lambda(t)$. Let $t_0 \geq 0$ be fixed. Let g be an eigenvector corresponding to the largest eigenvalue $\Lambda(t_0)$ of $P(t_0)AP(t_0)$. Thus we take g, satisfying:

$$|g| = 1, \qquad P(t_0)g = g, \qquad P(t_0)Ag = \Lambda(t_0)g.$$

For t close to t_0 let us consider the vector $P(t)g/|P(t)g|$. From the definition of $\Lambda(t)$ it follows that $\Lambda(t) \geq (AP(t)g, P(t)g/|P(t)g|^2$. Therefore, if $t - t_0 < 0$

$$\frac{\Lambda(t) - \Lambda(t_0)}{t - t_0} \leq \frac{\dfrac{(AP(t), P(t)g)}{|P(t)g|^2} - \dfrac{(AP(t_0)g, P(t_0)g)}{|P(t_0)g|^2}}{t - t_0}$$

Consequently,

$$D_l\Lambda(t)\bigg|_{t=t_0} \le \frac{d}{dt}\left[\frac{(AP(t)g, P(t)g)}{|P(t)g|^2}\right]\bigg|_{t=t_0} \tag{3.2}$$

where we denoted

$$D_l\Lambda(t)\bigg|_{t=t_0} = \limsup_{t\uparrow t_0} \frac{\Lambda(t)-\Lambda(t_0)}{t-t_0}.$$

In order to compute the right-hand side of (3.2) let us note that

$$\frac{d}{dt}|P(t)g|^2\bigg|_{t=t_0} = 0.$$

Indeed

$$\begin{aligned}\frac{d}{dt}|P(t)g|^2 &= -2(P(t)g, (I-P(t))A(t)P(t)g + P(t)A(t)^*(I-P(t)g)\\ &= -2(P(t)g, P(t)A(t)^*(I-P(t))g)\end{aligned}$$

and at $t = t_0$, $(I - P(t))g = 0$. Thus

$$\begin{aligned}\frac{d}{dt}\frac{(AP(t)g, P(t)g)}{|P(t)g|^2}\bigg|_{t=t_0} &= 2\frac{\left(AP(t)g, \left(\frac{d}{dt}P(t)\right)g\right)}{|P(t)g|^2}\bigg|_{t=t_0}\\ &= -2(Ag, (I-P(t_0))A(t_0)g)\\ &= -2(Ag - P(t_0)Ag, A(t_0)g)\\ &= -2((A-\Lambda(t_0))g, Ag) - 2((A-\Lambda(t_0))g, L(t_0)g).\end{aligned}$$

Now we note that

$$((A-\Lambda(t_0))g, Ag) = |(A-\Lambda(t_0))g|^2 \tag{3.3}$$

since $(g, Ag) = \Lambda(t_0)$. Therefore we obtain

$$D_l\Lambda(t)|_{t=t_0} \le -2|(A-\Lambda(t_0))g|^2 - 2((A-\Lambda(t_0))g, (I-P(t_0))L(t_0)g). \tag{3.4}$$

For $\lambda(t)$ we proceed in a similar manner. Let $h \in \mathscr{D}(A)$ satisfy $|h| = 1$, $P(t_0)h = 0$, $(I - P(t_0))Ah = \lambda(t_0)h$. (The existence of such an h is guaranteed by the fact that A^{-1} is compact.) Then

$$\underline{D}_l\lambda(t)\bigg|_{t=t_0} \ge \frac{d}{dt}\frac{(A(I-P(t))h; (I-P(t))h)}{|(I-P(t))h|^2}\bigg|_{t=t_0} \tag{3.5}$$

where

$$\underline{D}_l\lambda(t)\bigg|_{t=t_0} = \liminf_{t\uparrow t_0}\frac{\lambda(t)-\lambda(t_0)}{t-t_0}.$$

But

$$\frac{d}{dt}\frac{(A(I-P(t))h,(I-P(t))h)}{|(I-P(t))h|^2}\bigg|_{t=t_0}$$
$$= 2(Ah, P(t_0)A(t_0)^*h)$$
$$= 2P(t_0)Ah, A(t_0)^*h) = 2(Ah - \lambda(t_0)h, A(t_0)^*h).$$

We obtained

$$\underline{D}_l\lambda(t)|_{t=t_0} \geq 2|(A-\lambda(t_0))h|^2 + 2((A-\lambda(t_0))h, P(t_0)L(t_0)^*h). \tag{3.6}$$

Let us assume now that concerning $L(t)$ we have the estimates

$$|L(t)g| \leq [k_1|g|^2 + k_2|A^{1/4}g|^2 + k_3|A^{1/2}g|^2]^{1/2} \tag{3.7}$$
$$|L(t)^*g| \leq [k_1|g|^2 + k_2|A^{1/2}g|^2 + k_3|A^{1/2}g|^2]^{1/2} \tag{3.8}$$

for any $t \geq 0$, $g \in \mathscr{D}(A)$. Then from (3.4), (3.6) we obtain the estimates

$$\bar{D}_l\Lambda(t)|_{t=t_0} \leq -|(A-\Lambda(t_0))g|^2 + k_1 + k_2|A^{1/4}g|^2 + k_3|A^{1/2}g|^2 \tag{3.9}$$
$$\underline{D}_l\lambda(t)|_{t=t_0} \geq |(A-\lambda(t_0))h|^2 - k_1 - k_2|A^{1/4}h|^2 - k_3|A^{1/2}h|^2. \tag{3.10}$$

Now $|A^{1/4}g|^2 \leq |g||A^{1/2}g| = |A^{1/2}g|$ and $|A^{1/2}g|^2 = (Ag, g) = \Lambda(t_0)$.

Let us denote by $\sigma(A)$ the set of *distinct* eigenvalues of A, i.e., $\sigma(A) = \{\Lambda_k\}_{k=1,2,\ldots}$. Then, since $|g| = 1$,

$$|(A-\Lambda(t_0))g| \geq \operatorname{dist}(\Lambda(t_0), \sigma(A)), \qquad |(A-\lambda(t_0))h| \geq \operatorname{dist}(\lambda(t_0), \sigma(A)).$$

Finally, we obtain from (3.9), (3.10):

Theorem 3.2. *If* $\lambda(t) = \lambda(P(t))$, $\Lambda(t) = \Lambda(P(t))$ *are as defined in Definition* 3.1, *if* $L(t)$, $L(t)^*$ *satisfy* (3.7), (3.8), *and if* $P(t)$ *solve equation* (2.11), *then*

$$\bar{D}_l\Lambda(t) \leq -(\operatorname{dist}(\Lambda(t), \sigma(A)))^2 + k_1 + k_2(\Lambda(t))^{1/2} + k_3\Lambda(t), \tag{3.11}$$
$$\underline{D}_l\lambda(t) \geq (\operatorname{dist}(\lambda(t), \sigma(A)))^2 - k_1 - k_2(\lambda(t))^{1/2} - k_3\lambda(t). \tag{3.12}$$

Let us assume that $\sigma(A)$ has large gaps, namely that there exist m's such that

$$\left[\frac{\Lambda_{m+1}-\Lambda_m}{2}\right]^2 > k_1 + k_2\left[\frac{\Lambda_m+\Lambda_{m+1}}{2}\right]^{1/2} + k_3\frac{\Lambda_m+\Lambda_{m+1}}{2}. \tag{3.13}$$

Corollary 3.3 (Spectral Blocking Property).

(a) *If for some m satisfying* (3.13) *and* $t_0 \geq 0$

$$\Lambda(t_0) < \frac{\Lambda_m+\Lambda_{m+1}}{2} \quad \textit{then}$$

$$\Lambda(t) < \frac{\Lambda_m+\Lambda_{m+1}}{2} \quad \textit{for all } t \geq t_0.$$

(b) *If for some m satisfying* (3.13) *and* $t_0 \geq 0$

$$\lambda(t_0) > \frac{\Lambda_m + \Lambda_{m+1}}{2} \quad \textit{then}$$

$$\lambda(t) > \frac{\Lambda_m + \Lambda_{m+1}}{2} \quad \textit{for all } t \geq t_0.$$

Proof. From (3.13) and (3.11) (resp. (3.12)) it follows that

$$\bar{D}_l \Lambda(t)|_{\Lambda(t)=(\Lambda_m+\Lambda_{m+1})/2} < 0,$$

$$[\text{resp. } \underline{D}_l \lambda(t)|_{\lambda(t)=(\Lambda_m+\lambda_{m+1})/2} > 0]. \qquad \square$$

The condition (3.13) encountered as a technical assumption in works on inertial manifolds (see [FNST1]) is therefore a natural one: it requires the spectrum of A to have large enough gaps so that $\lambda(t)$ (resp. $\Lambda(t)$) cannot cross them from the right (resp. left).

CHAPTER 4

Strong Squeezing Property

Let $u_i(t) = S(t)u_i^0$, $i = 1, 2$, be two solutions of (2.1). Then their difference $w = u_1(t) - u_2(t)$ satisfies the equation

$$\frac{dw}{dt} + \mathscr{A}(t)w = 0, \tag{4.1}$$

$$w(0) = w_0 = u_1^0 - u_2^0, \tag{4.2}$$

where

$$\begin{aligned} \mathscr{A}(t)g &= Ag + Cg + B(\tilde{u}(t), g) + B(g, \tilde{u}(t)), \\ \tilde{u}(t) &= \tfrac{1}{2}(u_1(t) + u_2(t)). \end{aligned} \tag{4.3}$$

We denote by $\tilde{L}(t)$ the lower order terms of $\mathscr{A}(t)$

$$\tilde{L}(t)g = Cg + B(\tilde{u}(t), g) + B(g, \tilde{u}(t)). \tag{4.4}$$

Let us note that, for k_1, k_2, k_3 as in (3.7) (and (3.8)), $\tilde{L}(t)$ satisfies (3.7) (and (3.8)).

The strong squeezing property is an application of the spectral blocking property in the case of one-dimensional projectors. Thus, let $n = 1$ and $P(t) = (w(t) \otimes w(t))/|w(t)|^2$ be the one-dimensional projection on $w(t)/|w(t)|$:

$$P(t)g = \left(g, \frac{w(t)}{|w(t)|}\right)\frac{w(t)}{|w(t)|}.$$

Then $P(t)$ satisfies (2.11) with $A(t)$ replaced by $\mathscr{A}(t)$. The quantity $\Lambda(t)$ in this case equals $(Aw(t), w(t)/|w(t)|^2$. We can therefore apply Theorem 3.2 and Corollary 3.3 and obtain

Lemma 4.1. *If for some* $t_0 \geq 0$ *and* m *satisfying* (3.13)

$$\frac{(Aw(t_0), w(t_0))}{|w(t_0)|^2} < \frac{\Lambda_m + \Lambda_{m+1}}{2},$$

then for all $t \geq t_0$

$$\frac{(Aw(t), w(t))}{|w(t)|^2} < \frac{\Lambda_m + \Lambda_{m+1}}{2}.$$

We shall show that the only way in which $(Aw(t), w(t))/|w(t)|^2$ might be larger than or equal to $(\Lambda_m + \Lambda_{m+1})/2$ for all time is if $w(t)$ decays exponentially.

At this stage we must introduce two basic assumptions that are valid for practically all dissipative PDEs. Of course, these assumptions will be verified by the examples we will consider later, but they are also verified for the Navier–Stokes equations, as well as for the Boussinesq equations etc. The first assumption is a coercivity property for the linear part of the equation, that is,

$$(Ag + Cg, g) \geq k_4(|A^{1/2}g|^2 + |g|^2) \tag{4.5}$$

for all g in the domain $\mathscr{D}(A)$ of A and for some constant $k_4 > 0$. The second one is on the solutions $u(t)$, under investigation, as well as on the nonlinear part, namely

$$|(B(u(t), w) + B(w, u(t)), w)| \leq k_5|A^{1/2}w|^{1/2}|w|^{3/2} + k_6|w|^2 \tag{4.5a}$$

for all $t \geq 0$ and $w \in \mathscr{D}(A)$. Here k_5, k_6 are some constants independent of t, w, and $u(\cdot)$. Obviously in this case we also have

$$|(B(\tilde{u}(t), w) + B(w, \tilde{u}(t)), w)| \leq k_5|A^{1/2}w|^{1/2}|w|^{3/2} + k_6|w|^2 \tag{4.5b}$$

for all $t \geq 0$ and $w \in \mathscr{D}(A)$.

The conditions (4.5), (4.5a), (4.5b) are modeled on the Kuramoto–Sivashinsky equations and on the 2D Navier–Stokes equations. They are, however, general enough to display all aspects of our approach. In Chapter 17 we will illustrate the elasticity of our approach by treating a case (namely the Cahn–Hilliard equations) in which the above equations must be modified.

From (4.1) and (4.2) it follows that

$$\frac{1}{2}\frac{d}{dt}|w|^2 + k_4|A^{1/2}w|^2 + (B(\tilde{u}, w) + B(w, \tilde{u}, w) \leq 0,$$

and since from (4.5b) it follows

$$|(B(\tilde{u}, w) + B(w, \tilde{u})), w)| \leq \frac{k_4}{2}|A^{1/2}w|^2 + \frac{k_7}{2}|w|^2$$

where $k_7 = 2k_6 + \frac{3}{2}2^{-1/3}k_4^{-1/3}k_5^{4/3}$, we obtain

$$\frac{d}{dt}|w|^2 + (k_4\Lambda(t) - k_7)|w|^2 \leq 0$$

with

$$\Lambda(t) = \frac{|A^{1/2}w(t)|^2}{|w(t)|^2}.$$

Thus as long as $\Lambda(t) \geq (\Lambda_m + \Lambda_{m+1})/2$ we infer

$$\frac{d}{dt}|w(t)|^2 + \left[k_4 \frac{\Lambda_m + \Lambda_{m+1}}{2} - k_7\right]|w|^2 \leq 0.$$

Suppose now that m satisfying (3.13) is chosen large enough to satisfy also

$$\frac{\Lambda_m + \Lambda_{m+1}}{4} > \frac{k_7}{k_4} \tag{4.6}$$

and that

$$\Lambda(0) = \frac{(Aw_0, w_0)}{|w_0|^2} \geq \frac{\Lambda_m + \Lambda_{m+1}}{2}.$$

Let $T = \sup\{t \geq 0 | \Lambda(t) \geq (\Lambda_m + \Lambda_{m+1})/2\}$. If $T = 0$ then for all $t > 0$ it follows that $\Lambda(t) < (\Lambda_m + \Lambda_{m+1})/2$. If $0 < T \leq \infty$, Lemma 4.1 implies that $\Lambda(t) \geq (\Lambda_m + \Lambda_{m+1})/2$ for all $0 \leq t \leq T$, so it follows that

$$|w(t)|^2 \leq |w(0)|^2 \exp\left[-k_4 \frac{\Lambda_m + \Lambda_{m+1}}{2} + k_7\right]t, \qquad 0 \leq t \leq T. \tag{4.7}$$

Let us choose n such that

$$5(\Lambda_m + \Lambda_{m+1}) < \lambda_{n+1}. \tag{4.8}$$

Then if

$$\frac{(Aw, w)}{|w|^2} < \frac{\Lambda_m + \Lambda_{m+1}}{2}$$

it follows that

$$|Q_n w| \leq \frac{|P_n w|}{3}$$

where P_n is the projection on the span of $\{w_1, \ldots, w_n\}$ and $Q_n = I - P_n$. Indeed

$$\frac{\Lambda_m + \Lambda_{m+1}}{2} > \frac{(Aw, w)}{|w|^2} \geq \frac{\lambda_{n+1}|Q_n w|^2}{|P_n w|^2 + |Q_n w|^2} = \frac{\lambda_{n+1}}{[|P_n w|^2/|Q_n w|] + 1}.$$

We proved therefore

Theorem 4.2 (Strong Squeezing Property). *Let n satisfy* (4.8) *where m satisfies* (4.6) *and* (3.13). *Then either*

(a) $|w(t)| \leq |w(0)| \exp(-k_8 t)$ for all $t > 0$, where $k_8 = \dfrac{1}{2}\left(k_4 \dfrac{\Lambda_m + \Lambda_{m+1}}{2} - k_7\right)$,

or

(b) *there exists $t_0 > 0$ such that the inequality in* (a) *is valid for $t < t_0$ and $w(t) \in \left\{w \in H | (Aw, w) \leq \dfrac{\Lambda_m + \Lambda_{m+1}}{2}|w|^2\right\} \subset \{w \in H | |(I - P_n)w| \leq \frac{1}{3}|P_n w|\}$ for all $t \geq t_0$.*

Replacing $\tilde{u}(t)$ with $u(t) = S(t)u_0$ in the arguments leading to Theorem 4.2, we obtain

Theorem 4.3. *Let n be chosen as in Theorem* 4.2. *Let $w(t)$ be any solution of* (2.3), *the linearized equation along $S(t)u_0$. Then the conclusions of Theorem* 4.2 *hold for $w(t)$.*

CHAPTER 5

Cone Invariance Properties

One of the features of a dissipative equation of type (2.1), (2.2) is the existence of compact absorbing sets. More precisely, there exists $Y \subset H$ satisfying

Y is convex, closed in H, a bounded neighborhood of 0 in $\mathscr{D}(A^{1/4})$ (in particular, Y is compact in H). (5.1)

For every $\theta \geq 1$ and any $u_0 \in \theta Y$ the inequalities (3.7), (3.8), and (4.5a) are valid. The constants k_1, k_2, k_3, k_5, and k_6 depend on θ only. (5.2)

The set Y is absorbing; i.e., if Z is any bounded set in H, there exists a $t_0 \geq 0$ (depending on Z) such that $S(t)Z \subset Y$ for $t \geq t_0$. (5.3)

In applications it can occur that the convex closed sets Y in H vested with properties (5.2), (5.3) are bounded in some $\mathscr{D}(A^\alpha)$ with $\frac{1}{4} < \alpha \leq 1$. In that case, the role of the power $A^{1/4}$ has to be played by the power A^α. The necessary modifications in the sequel are transparent enough (see Chapters 16 and 17 for illustration).

Let m be an integer. Let us consider the cones in H

$$C_{m,\gamma} = \{w \in H \,|\, |(I - P_m)w| \leq \gamma |P_m w|\}. \tag{5.4}$$

We shall prove in this section invariance properties of these cones with respect to the equations (2.3) and (4.1) of evolution of infinitesimal and finite displacements, respectively. We shall make the assumption

$$|(L(t)w, w)| \leq k_9 |A^{1/4} w|^2 + k_9' |w|^2 \tag{5.5}$$

valid for all $w \in \mathscr{D}(A^{1/2})$, $t \geq 0$. The constants k_9 and k_9' are uniform for all $u_0 \in \theta Y$, for each fixed $\theta \geq 1$.

Let w be a solution of (4.1), (4.2) such that $u_1^0, u_2^0 \in \theta Y$ ($\theta \geq 1$). Let us write $w(t) = p(t) + q(t)$ where $p(t) = P_m w(t)$ and $q(t) = (I - P_m)w(t)$, $m \geq 1$. Let

$\gamma > 0$ be fixed. The equation of evolution of $\delta(t) = |q(t)|^2 - \gamma^2|p(t)|^2$ is obtained by taking the scalar product of (4.1) with $q(t) - \gamma^2 p(t)$:

$$\frac{1}{2}\frac{d}{dt}\delta(t) + |A^{1/2}q|^2 - \gamma^2|A^{1/2}p|^2 + (\tilde{L}(t)w, q - \gamma^2 p) = 0.$$

Assumption (5.5), which is valid for $\tilde{L}(t)$ too, enables us to estimate

$$\begin{aligned}
&|(\tilde{L}(t)w, q - \gamma^2 p)| \\
&= |(\tilde{L}(t)q, q) - \gamma^2(\tilde{L}(t)p, p) - \gamma^2(\tilde{L}(t)q, p) + (\tilde{L}(t)p, q)| \\
&\le k_9|A^{1/4}q|^2 + k_9'|w|^2 + |q|(|\tilde{L}(t)p| + \gamma^2|\tilde{L}(t)^*p|) + \gamma^2|\tilde{L}(t)p||p| \\
&\le k_9|A^{1/4}q|^2 + k_9'|w|^2 + (1 + \gamma^2)|q|(k_1|p|^2 + k_2|A^{1/4}p|^2 + |A^{1/2}p|^2)^{1/2} \\
&+ \gamma^2|p|(k_1|p|^2 + k_2|A^{1/4}p|^2 + k_3|A^{1/2}p|^2)^{1/2}.
\end{aligned}$$

Since $p \in P_m H$ we have

$$k_1|p|^2 + k_2|A^{1/4}p|^2 + k_3|A^{1/2}p|^2 \le (k_1' + k_2\lambda_m^{1/2} + k_3\lambda_m)|p|^2,$$

where $k_1' = k_1 + k_9'^2$. If $|q| = \gamma|p|$ then

$$|(\tilde{L}(t)w, q - \gamma^2 p)| \le k_9|A^{1/4}q|^2 + (\gamma + \gamma^2 + \gamma^3)(k_1' + k_2\lambda_m^{1/2} + k_3\lambda_m)^{1/2}|p|^2.$$

On the other hand, the term $|A^{1/2}q|^2 - \gamma^2|A^{1/2}p|^2$ can be estimated as follows:

$$|A^{1/2}q|^2 - \gamma^2|A^{1/2}p|^2 \ge \lambda_{m+1}^{1/2}|A^{1/4}q|^2 - \gamma^2\lambda_m|p|^2.$$

We thus obtain an estimate of $\frac{1}{2}d(\delta(t))/dt$ at a point where $\delta(t) = 0$ (i.e., $|q| = \gamma|p|$):

$$\begin{aligned}
\frac{1}{2}\frac{d}{dt}\delta(t)\bigg|_{\delta=0} \le &-[(\lambda_{m+1}^{1/2} - k_9)|A^{1/4}q|^2 - \gamma^2\lambda_m|p|^2 \\
&- (\gamma + \gamma^2 + \gamma^3)(k_1' + k_2\lambda_m^{1/2} + k_3\lambda_m)^{1/2}|p|^2].
\end{aligned}$$

Now $(\lambda_{m+1}^{1/2} - k_9)|A^{1/4}q|^2 \ge (\lambda_{m+1}^{1/2} - k_9)\lambda_{m+1}^{1/2}|q|^2$ if $\lambda_{m+1}^{1/2} \ge k_9$. We deduce that $\frac{1}{2}d(\delta(t))/dt|_{\delta=0} \le 0$ if

$$\gamma^2[(\lambda_{m+1}^{1/2} - k_9)\lambda_{m+1}^{1/2} - \lambda_m] - (\gamma + \gamma^2 + \gamma^3)(k_1' + k_2\lambda_m^{1/2} + k_3\lambda_m)^{1/2} \ge 0.$$

This last inequality will be true if

$$\lambda_{m+1} - \lambda_m \ge k_9\lambda_m^{1/2} + \frac{1 + \gamma + \gamma^2}{\gamma}(k_1' + k_2\lambda_m^{1/2} + k_3\lambda_m)^{1/2}, \tag{5.6}$$

where $k_1' = k_1 + k_9'^2$ and k_9, k_9' were introduced in (5.5). This inequality is entirely similar to (3.13). If (5.6) is satisfied it follows that if $w_0 \in C_{m,\gamma}$, i.e., if $\delta(0) \le 0$, then $\delta(t) \le 0$, i.e., $w(t) \in C_{m,\gamma}$ for all $t \ge 0$.

We proved therefore Theorem 5.1.

Theorem 5.1 (The Cone Invariance Property). *Let* $\gamma > 0$, $\theta \ge 1$ *be fixed. Let* $m \ge 1$ *be such that* (5.6) *is satisfied. Consider* $w(t) = u_1(t) - u_2(t)$ *the difference* *of the solutions* $u_1(t)$, $u_2(t)$ *of* (2.1) *with initial data* $u_1(0) \in \theta Y$, $u_2(0) \in \theta Y$. *Then if* $w(0) \in C_{m,\gamma}$ *it follows that* $w(t) \in C_{m,\gamma}$ *for all* $t \geq 0$.

In an entirely similar manner one proves Theorem 5.2.

Theorem 5.2. *Let* $\gamma > 0, \theta \geq 1$ *be fixed. Let* $m \geq 1$ *be such that* (5.6) *is satisfied. Denote by* $v(t)$ *a solution of* (2.3) *with* $u_0 \in \theta Y$ *starting at* $t = 0$ *from* $v(0) \in C_{m,\gamma}$. *Then* $v(t) \in C_{m,\gamma}$ *for all* $t \geq 0$.

Remark 5.3. It is easy to show that if in addition to (5.6) the condition

$$\lambda_{n+1} \geq \frac{2(1+\gamma^2)}{k_4\gamma^2}\left[k_9 + \frac{(k_9')^2}{2k_4}\right] \tag{5.6a}$$

is satisfied, then in Theorems 5.1 and 5.2 one can add the same kind of alternative as the one to be found in the statement of Theorem 4.2: Either

(a) $|w(t)| \leq |w(0)| \exp(-k_8' t)$ for all $t > 0$

or

(b) there exists $t_0 > 0$ such that the inequality in (a) holds for $t < t_0$ and $w(t) \in C_{n,\gamma}$ for $t \geq t_0$.

Let us consider the locally compact cones

$$K_n = \left\{ w \in \mathscr{D}(A^{1/2}) \,|\, |A^{1/2}w|^2 \leq \frac{\lambda_n + \lambda_{n+1}}{2}|w|^2 \right\}. \tag{5.7}$$

A simple computation (see Chapter 4 for the case $\gamma = \frac{1}{3}$) shows that

$$K_n \subset C_{l,\gamma} \quad \text{if} \quad \frac{\lambda_n + \lambda_{n+1}}{2} \leq \frac{\gamma^2}{1+\gamma^2}\lambda_{l+1}. \tag{5.8}$$

Now Lemma 4.1 and its analog for equation (2.3) can be reformulated as cone invariance properties. These properties, although they are valid in what appear to be more general circumstances (assumption (5.5) is not needed for them), are in a sense much stronger than Theorem 5.1 and Theorem 5.2 because of (5.8) and the fact that K_n are locally compact, respectively. We conclude this section with the precise statements of these locally compact cone invariance properties.

Theorem 5.4. *Let* $\theta \geq 1$. *Let* λ_n *satisfy*

$$\left[\frac{\lambda_{n+1} - \lambda_n}{2}\right]^2 > k_1 + k_2\left[\frac{\lambda_n + \lambda_{n+1}}{2}\right]^2 + k_3\frac{\lambda_n + \lambda_{n+1}}{2}. \tag{5.9}$$

Let $w(t) = u_1(t) - u_2(t)$ *be the difference of two solutions* $u_1(t)$, $u_2(t)$ *of* (2.1) *with initial data* $u_1(0) \in \theta Y$, $u_2(0) \in \theta Y$. *Then* $u_1(0) - u_2(0) \in K_n$ *implies* $u_1(t) - u_2(t) \in K_n$ *for all* $t \geq 0$.

Theorem 5.5. *Let $\theta \geq 1$. Let λ_n satisfy* (5.9). *Let $w(t)$ be a solution of* (2.3) *with $u_0 \in \theta Y$ and $w(0) \in K_n$. Then $w(t) \in K_n$ for all $t \geq 0$.*

Let us comment on the meaning of (5.9). The constants k_1, k_2, k_3 are those which are given uniformly in (5.2). The inequality (5.9) is equivalent to the fact that $\lambda_n = \Lambda_m$ for some m satisfying (3.13).

CHAPTER 6

Consequences Regarding the Global Attractor

Let X be the global attractor of the dissipative system under consideration. Recall that X is the largest set in H with the properties

(i) $S(t)X = X$ for $t \geq 0$,
(ii) X is bounded in H,
(iii) $\mathrm{dist}(S(t)u_0, X) \to 0$ as $t \to \infty$ for all $u_0 \in H$.

This set is actually given by the formula

$$X = \bigcap_{t \geq 0} \left(\overline{\bigcup_{s \geq t} S(s)Y} \right)$$

where the closure is taken in H. Obviously X is nonempty and compact in H.

Let u_1, u_2 be two distinct points in the universal attractor X. Let P_n be a spectral projector for which Theorem 4.2 holds. We claim that $u_1 - u_2$ cannot belong to the set

$$C^c = \{w \mid |(I - P_n)w| > \tfrac{1}{3}|P_n w|\}.$$

Indeed, arguing by contradiction, assume that $u_1 - u_2$ belongs to C^c. Since $X = S(t)X$ for all $t > 0$, there exist v_1, v_2 in X such that $u_1 = S(t)v_1$, $u_2 = S(t)v_2$. By virtue of Theorem 4.2

$$|u_1 - u_2| \leq |S(t)v_1 - S(t)v_2| \leq |v_1 - v_2| \exp(-k_8 t) \leq d \cdot \exp(-k_8 t),$$

where $d = \sup_{v, w \in X} |v - w|$. Since $t > 0$ is arbitrary, it follows $u_1 = u_2$, contradiction. We proved therefore

Theorem 6.1. *Let n be chosen as in Theorem 4.2. Then for every u_1, u_2 in X*

$$|(I - P_n)(u_1 - u_2)| \leq \tfrac{1}{3}|P_n(u_1 - u_2)|. \tag{6.1}$$

In particular, P_n is injective when restricted to X and its inverse is Lipschitz.

Remark 6.1a. Let m be as above (i.e., satisfying (3.13)) and let $\lambda_N = \Lambda_m$, $\lambda_{N+1} = \Lambda_{m+1}$. Then P_n restricted to X is injective and its inverse is Lipschitz, but the Lipschitz constant may be larger than $\frac{4}{3}$.

Indeed, by replacing Theorem 4.2 with Theorem 5.3 in the argument leading to Theorem 6.1 we can obtain easily that

$$(A(u_1 - u_2), u_1 - u_2) \leq \frac{\Lambda_m + \Lambda_{m+1}}{2} |u_1 - u_2|^2 \tag{6.1a}$$

for all $u_1, u_2 \in X$. This implies

$$|(I - P_N)(u_1 - u_2)|^2 \leq \frac{\lambda_{N+1} + \lambda_N}{\lambda_{N+1} - \lambda_N} |P_n(u_1 - u_2)|^2, \tag{6.1b}$$

for all $u_1, u_2 \in X$.

If, however, n satisfies solely the requirements (5.6) and (5.6a) for $\gamma = \frac{1}{3}$, then (6.1) is true, as one can easily see using Theorem 5.1 and Remark 5.3.

An argument similar to the one used above shows that as long as a solution $S(t)u_0$, for some $u_0 \in H$, stays in the region

$$\bigcup_{x \in X} \{u \mid |(I - P_n)(u - x)| > \tfrac{1}{3}|P_n(u - x)|\}$$

its distance to X decreases exponentially:

$$\operatorname{dist}(S(t)(u_0, X) \leq \left[\max_{u \in X} |u_0 - u|\right] \exp(-k_8 t).$$

Therefore the complementary region is of particular interest. Let us denote it by $C_{n,X}$:

$$C_{n,X} = \{u \in H \mid |(I - P_n)(u - x)| \leq \tfrac{1}{3}|P_n(u - x)| \quad \text{for } every\ x \in X\}. \tag{6.2}$$

Now, clearly $C_{n,X}$ has the following invariance under $S(t)$: $S(t)(C_{n,X} \cap Y) \subset C_{n,X}$ for $t \geq 0$. By Theorem 6.1, $X \subset C_{n,X}$. The strong squeezing property readily yields the following property of the attractor.

Theorem 6.2. *Let P_n be as in Theorem 4.2 and Theorem 5.1 and let $C_{n,X}$ be the set defined in (6.2). Then $S(t)(C_{n,X} \cap Y) \subset C_{n,X}$ for all $t \geq 0$. Moreover, for an arbitrary solution $S(t)u_0$, $u_0 \in Y$, either*

(a) $\operatorname{dist}(S(t)u_0, X) \leq [\max_{u \in X} |u_0 - u|] \exp(-k_8 t)$ *for all* $t > 0$

or

(b) *the inequality in (a) holds up to a finite time $t_0 \geq 0$ and, for all $t \geq t_0$, $S(t)u_0$ belongs to $C_{n,X}$.*

We conclude this chapter by showing that the complement of a large ball in $P_n H$ is included in $C_{n,X}$. Indeed, let ρ_0 and ρ_1 be defined as

$$\rho_0 = \sup_{v \in X} |v|, \qquad \rho_1 = \max_{v \in X} |A^{1/4} v|.$$

Obviously $\rho_0 \le \lambda_1^{-1/4}\rho_1 < \infty$ by $X \subset Y$. Then for $x \in X$ and $u \in P_n H$ we have

$$|(I - P_n)(u - x)| \le \frac{1}{\lambda_{n+1}^{1/4}} |A^{1/4}(I - P_n)(u - x)| \le \frac{\rho_1}{\lambda_{n+1}^{1/4}}$$

and

$$|u| - \rho_0 \le |u - x| \le |P_n(u - x)| + \frac{\rho_1}{\lambda_{n+1}^{1/4}}; \tag{6.3}$$

thus, if $|u| - \rho_0 \ge 4\rho_1/\lambda_{n+1}^{1/4}$ then $|(I - P_n)u| \le \frac{1}{3}|P_n u|$. Therefore we deduce Lemma 6.3.

Lemma 6.3. *Let* $C_{n,X}$ *be defined as in* (6.2). *Then*

$$\{u | u = P_n u, |u| = R\} \subset C_{n,X}$$

provided

$$R \ge \rho_0 + \frac{4\rho_1}{\lambda_{n+1}^{1/4}}. \tag{6.4}$$

Let us mention that there is nothing special about the number 3 in the condition $|(I - P_n)w| \le \frac{1}{3}|P_n w|$. The $\frac{1}{3}$ coefficient can be made arbitrarily small by taking larger n; our choice is motivated by the need to apply some triangle inequalities in Chapter 10.

CHAPTER 7

Local Exponential Decay Toward Blocked Integral Surfaces

Suppose Σ is an n-dimensional integral surface in Y, that is, an n-dimensional manifold without boundary that is positively invariant. Let, for each $u \in \Sigma$, $P(u)$ denote the projector on the tangent space $T_u(\Sigma)$ to Σ at u. Let us assume that the surface is blocked in the sense that

$$\lambda(P(u)) > \frac{\lambda_n + \lambda_{n+1}}{2} \quad \text{for all } u \tag{7.1}$$

and that $\lambda_n = \Lambda_m$ which satisfies condition (3.13). Let us consider $u_0 \in H$ and assume that the distance between u_0 and Σ is attained at some $u_1 \in \Sigma$. Then, clearly $P(u_1)(u_0 - u_1) = 0$. Let us consider the trajectories $S(t)u_0$, $S(t)u_1$. Their difference $w(t) = S(t)u_0 - S(t)u_1$ satisfies (4.1). Denoting $\Lambda(t) = (Aw(t), w(t))/|w(t)|^2$, we have as in Chapter 4:

$$\frac{d}{dt}|w(t)|^2 + (k_4\Lambda(t) - k_7)|w|^2 \leq 0. \tag{7.2}$$

We note that for $t = 0$

$$\Lambda(0) = \frac{(Aw(0), w(0))}{|w(0)|^2} \geq \lambda(P(u_1)) > \frac{\Lambda_n + \Lambda_{n+1}}{2},$$

since $P(u_1)w_0 = 0$. Therefore, reading (7.2) for $t = 0$ we conclude that

$$\frac{d}{dt}|w(t)|^2\bigg|_{t=0} \leq -\left[k_4\frac{\Lambda_n + \Lambda_{n+1}}{2} - k_7\right]|w(0)|^2,$$

where the derivative is only the right-sided one. Now, for $t > 0$,

$$(\mathrm{dist}(S(t)u_0, \Sigma))^2 \leq |w(t)|^2 \quad \text{and therefore}$$

$$D_r(\mathrm{dist}(S(t)u_0, \Sigma))^2|_{t=0} \leq -\left[k_4 \frac{\Lambda_n + \Lambda_{n+1}}{2} - k_7\right](\mathrm{dist}(u_0, \Sigma))^2.$$

We summarize as follows:

Lemma 7.1. *Let $\Sigma \subset Y$ be an n-dimensional integral surface satisfying* (7.1). *Let $u_0 \in H$. Then, as long as* $\mathrm{dist}(S(t)u_0, \Sigma)$ *is attained on* Σ,

$$\mathrm{dist}(S(t)u_0, \Sigma)^2 \leq \mathrm{dist}(u_0, \Sigma)^2 \exp\left[-\left(k_4 \frac{\Lambda_n + \Lambda_{n+1}}{2} - k_7\right)t\right]. \tag{7.3}$$

In the sequel we shall use Lemma 7.1 for n large enough to ensure (4.8), (4.6), and (3.13). Also we will, if necessary, replace Y with some θY, with $\theta > 1$, since we have chosen $\theta = 1$ in Chapters 6 and 7 only for convenience.

CHAPTER 8

Exponential Decay of Volume Elements and the Dimension of the Global Attractor

Let Σ_0 be an m-dimensional smooth manifold in θY for some fixed $\theta \in [1, \infty)$, let $u_0 \in \Sigma_0$, and let $u = \varphi(\alpha)$ be a local parametrization of Σ_0 in a neighborhood of u_0, where $\alpha = (\alpha_1, \dots, \alpha_m)$ runs over a neighborhood of 0 in $\mathbb{R}^m$ and $u_0 = \varphi(0)$. The infinitesimal volume element of $S(t)\Sigma_0$ at $S(t)u_0$ is $|v_1(t) \wedge \cdots \wedge v_m(t)|$ where $v_i(t)$ evolve according to (2.3) and $v_i(0) = \partial\varphi(\alpha)/\partial\alpha_i|_{\alpha=0}$. Using (2.7) and (2.9) we deduce the equation (see [CF1])

$$\frac{1}{2}\frac{d}{dt}|v_1(t) \wedge \cdots \wedge v_m(t)|^2 + (\operatorname{Tr} A(t)P(t))|v_1(t) \wedge \cdots \wedge v_m(t)|^2 = 0, \quad (8.1)$$

where $P(t)$ is the projector on the tangent space to $S(t)\Sigma_0$ at $S(t)u_0$. Thus the volume element will decay exponentially if

$$\liminf_{t\to\infty} \frac{1}{t}\int_0^t \operatorname{Tr}(A(s)P(s))\,ds > 0. \quad (8.2)$$

In order to give an explicit sufficient condition for (8.2) to hold, let us make the following definition: For $l = 1, 2, \dots$

$$q_l = \liminf_{t\to\infty} \frac{1}{t}\left[\inf_{u_0} \int_0^t \inf_Q \operatorname{Tr}(A(s)Q)\,ds\right] \quad (8.3)$$

where the first infimum is taken over all initial data $u_0 \in Y$, $A(s)$ this time is the linearization of $N(\cdot)$ at $S(s)u_0$, and the second infimum is taken over all orthogonal projections in H with range in $\mathcal{D}(A)$ and of rank l. Obviously, (8.2) holds proved $q_l > 0$.

Lemma 8.1. *For* $l = 1, 2, \dots$ *we have*

$$q_l \geq \bar{q}_l = \tfrac{1}{2}(\lambda_1 + \cdots + \lambda_l) - [\tfrac{1}{2}k_9^2 + k_9']l. \quad (8.4)$$

PROOF. Let $\{\phi_1, \dots, \phi_l\} \subset \mathscr{D}(A)$ be an orthonormal system in H, let Q be the orthogonal projection on the space spanned by $\phi_1, \dots, \phi_l$, and let $A(s)$ be as in (8.3). Then, by (5.5),

$$\begin{aligned} \operatorname{Tr} A(s)Q &= \sum_1^l (A\phi_j, \phi_j) + \sum_1^l (L(s)\phi_j, \phi_j) \\ &\geq \sum_1^l (A\phi_j, \phi_j) - k_9' l - k_9 \sum_1^l |A^{1/4}\phi_j|^2 \\ &\geq \sum_1^l (A\phi_j, \phi_j) - k_9' l - k_9 \sum_1^l |A^{1/2}\phi_j| \\ &\geq \sum_1^l (A\phi_j, \phi_j) - k_9' l - k_9 l^{1/2} \left(\sum_1^l (A\phi_j, \phi_j) \right)^{1/2} \\ &\geq \frac{1}{2} \sum_1^l (A\phi_j, \phi_j) - k_9' l - \tfrac{1}{2} k_9^2 l \\ &\geq \tfrac{1}{2}(\lambda_1 + \cdots + \lambda_l) - k_9' l - \tfrac{1}{2} k_9^2 l \end{aligned} \tag{8.5}$$

(see [CFT], p. 47, for the justification of this last inequality). Now (8.4) follows readily from the definition (8.3). □

In almost all interesting examples of dissipative partial differential equations or systems of equations, the linear principal part is an elliptic operator of order at least 2. Therefore, if the dimension d of the space variable is ≤ 3 one has $\lambda_m \gtrsim m^{2/d}$ (see for instance [Ag, Met]); hence

$$\frac{\lambda_1 + \cdots + \lambda_l}{l} \to \infty \quad \text{for } l \to \infty. \tag{8.6}$$

From now on we shall assume that condition (8.6) is also satisfied by our operator A. Under this assumption let us define m_0 and m_1 as the first l's for which

$$\lambda_1 + \cdots + \lambda_l > (k_9^2 + 2k_9')l \quad \text{and} \quad \lambda_1 + \cdots + \lambda_l > 2(k_9^2 + 2k_9')l,$$

respectively (obviously $m_0 \leq m_1$). Then we have the following consequences of Lemma 8.1.

Corollary 8.2. *Let $\lambda_1 + \cdots + \lambda_m > (k_9^2 + 2k_9')m$ and let $\Sigma_0 \subset Y$ be an m-dimensional surface with finite volume $\operatorname{vol}_m(\Sigma_0)$. Then*

$$\operatorname{vol}_m(S(t)\Sigma_0) \leq e^{-k_{10}t} \operatorname{vol}_m(\Sigma_0), \qquad t > 0$$

where k_{10} is a constant independent of Σ_0.

PROOF. Without loss of generality we can asssume that Σ_0 has a parametrization φ as described at the beginning of this section. Let $v_i(t, \alpha)$ be the solution of (2.3), $v_{i\alpha}(0) = \partial\varphi/\partial\alpha_i$. Then

$$\begin{aligned}\mathrm{vol}_m(S(t)\Sigma_0) &= \int |v_1(t,\alpha) \wedge \cdots \wedge v_m(t,\alpha)|\, d\alpha \\ &\leq \int |\varphi'_{\alpha_1}(\alpha) \wedge \cdots \wedge \varphi'_{\alpha_m}| e^{-\bar{q}_m t}\, d\alpha = e^{-\bar{q}_m t}\, \mathrm{vol}_m(\Sigma_0) \\ &\leq e^{-\frac{1}{2}(\lambda_1 + \cdots + \lambda_m - k''_9 m)t}\, \mathrm{vol}_m(\Sigma_0)\end{aligned}$$

where $k''_9 = k_9^2 + 2k'_9$. □

Corollary 8.3. *Let Γ be an $(n-1)$-dimensional smooth surface in $Y \subset \mathscr{D}(A)$ with finite $\mathrm{vol}_{n-1}(\Gamma) < \infty$ and such that $\sup\{|N(u)| : u \in \Gamma\} < \infty$. Let Σ be the integral surface with initial data Γ*

$$\Sigma = \bigcup_{t>0} S(t)\Gamma.$$

then $\mathrm{vol}_n(\Sigma) < \infty$ provided $2\bar{q}_n = \lambda_1 + \cdots + \lambda_n - (k_9^2 + 2k'_9)n > 0$.

PROOF. Without loss of generality we can assume that Γ is parametrized by $\phi(\alpha)$ where α runs in an open bounded set $U \subset \mathbb{R}^{n-1}$. Then $S(t)\phi(\alpha)$ for $(t,\alpha) \in \mathbb{R}_+ \times U$ is a parametrization of Σ. Let $v_1(t,\alpha), \ldots, v_n(t,\alpha)$ be the solutions of (2.3) with initial data $\phi'_{\alpha_1}(\alpha), \ldots, \phi'_{\alpha_{n-1}}(\alpha), -N(\phi(\alpha))$, respectively. Then

$$\begin{aligned}\mathrm{vol}_n(\Sigma) &= \int_0^\infty \int_U |v_1(t,\alpha) \wedge \cdots \wedge v_n(t,\alpha)|\, dt\, d\alpha \\ &\leq \int_0^\infty e^{-\bar{q}_n t} \int_U |v_1(0,\alpha) \wedge \cdots \wedge v_n(0,\alpha)|\, d\alpha\, dt \\ &= \frac{1}{\bar{q}_n} \int_U |\phi'_{\alpha_1} \wedge \cdots \wedge \phi'_{\alpha_{n-1}} \wedge N(\phi(\alpha))|\, d\alpha' \\ &\leq \frac{1}{\bar{q}_n} (\max\{|N(u)| : u \in \Gamma\}) \cdot \mathrm{vol}_{n-1}(\Gamma).\end{aligned}$$

□

Theorem 8.4. *The Hausdorff dimension $d_H(X)$ and the fractal dimension $d_M(X)$ of the universal attractor X satisfy*

$$d_H(X) \leq m_0, \tag{8.7}$$

$$d_M(X) \leq m_0 + m_1. \tag{8.8}$$

Before giving the proof of this theorem let us recall the definition of these classical dimensions. Namely, $d_H(X)$, respectively $d_M(X)$, is the infimum of $d \in [0,\infty]$ such that for every $\delta, \varepsilon > 0$ there exists a covering of X by balls $\{B_\alpha\}_\alpha$ of radii $r_\alpha \leq \varepsilon$ satisfying

$$\sum r_\alpha^d \leq \alpha, \quad \text{respectively } \varepsilon^d \cdot (\#\alpha) < \delta.$$

PROOF OF THEOREM 8.4. We recall that according to [CFT], Theorem 4.1, we have that $q_m > 0$ implies $d_H(X) \leq m$ and

$$d_M(X) \leq m \cdot \max_{1 \leq l \leq m} \left[1 + \frac{-q_l}{q_m} \right].$$

For $m = m_0$ this directly yields (8.7). For $m = m_1$ we obviously have

$$\begin{aligned}\max_{1 \leq l \leq m_1} \left[1 + \frac{-q_l}{q_{m_1}} \right] \leq \max_{1 \leq l \leq m_0} \left\{ 1 + \frac{-q_l}{q_{m_1}} \right\} &\leq 1 \\ &+ \frac{(k_9^2 + 2k_9')m_0}{\lambda_1 + \cdots + \lambda_{m_1} - (k_9^2 + 2k_9')m_1} \\ &= 1 + \frac{m_0}{m_1}.\end{aligned}$$

□

CHAPTER 9

Choice of the Initial Manifold

We are now going to describe the set of initial data for our integral manifold. In order to make our treatment more transparent we shall assume that

$$(B(u,u),u) = 0 \quad \text{for all } u \in \mathscr{D}(A). \tag{9.1}$$

This orthogonality condition is valid for many dissipative partial equations (see Chapters 15 and 16). At the end of this chapter we shall briefly discuss the necessary changes in case (9.1) fails.

Let us choose n such that $\lambda_n < \lambda_{n+1}$. The assumption (9.1) allows us to take Γ to be a large sphere in $P_n H$, $\Gamma = \{u \mid P_n u = u, |u| = R\}$, R to be specified. In the absence of this assumption, as we shall show later, the geometry of the appropriate Γ will be less simple. We recall that P_n is the orthogonal projection in H onto the space spanned by the eigenvectors of A corresponding to eigenvalues $\lambda \leq \lambda_n$. At each $u \in \Gamma$ we consider the linear space $T_u(\Gamma) + (N(U))\mathbb{R}$ where $T_u(\Gamma)$ is the tangent hyperplane (in $P_n H$) to Γ at u and $N(u) = Au + R(u)$. We denote by $P(u)$ the orthogonal projection in H on this space and $\lambda(u) = \lambda(P(u))$, $\Lambda(P(u))$. The constraints that we wish to impose on Γ are the following:

(I) $\Lambda(u) < \dfrac{\lambda_n + \lambda_{n+1}}{2}$ for every $u \in \Gamma$,

(II) $\lambda(u) > \dfrac{\lambda_n + \lambda_{n+1}}{2}$ for every $u \in \Gamma$,

(III) $(N(u), u) > 0$.

We shall check that the constraints (I) to (III) can be satisfied provided n and R are large enough. We start with (III). Let us recall that $A + C$ is coercive (see (4.5)). Therefore,

$$\begin{aligned}(N(u), u) &= (Au + R(u), u) = (Au + B(u, u) + Cu + f, u)\\ &= (Au + Cu, u) + (f, u) \geq k_4[|A^{1/2}u|^2 + |u|^2] - |(f, u)|.\end{aligned}$$

For later convenience and without loss of generality we can take f of the form

$$f = A\varphi + \psi \tag{9.2}$$

and express our bounds in terms of φ and ψ. Thus

$$\begin{aligned}|(f, u)| &= |(A^{1/2}\varphi, A^{1/2}u) + (\psi, u)| \leq |A^{1/2}\varphi||A^{1/2}u| + |\psi||u|\\ &\leq \frac{k_4}{4}|A^{1/2}u|^2 + \frac{k_4}{4}|u|^2 + \frac{1}{k_4}[|A^{1/2}\varphi|^2 + |\psi|^2].\end{aligned}$$

Therefore

$$\begin{aligned}(N(u), u) &\geq \frac{k_4}{2}[|A^{1/2}u|^2 + |u|^2] \quad \text{if}\\ |A^{1/2}u|^2 + |u|^2 &\geq \frac{4}{k_4^2}[|A^{1/2}\varphi|^2 + |\psi|^2].\end{aligned} \tag{9.3}$$

In particular, (9.3) will be true if R satisfies

$$R \geq \frac{2}{k_4}\sqrt{|A^{1/2}\varphi|^2 + |\psi|^2} \tag{9.4}$$

where k_4 is the coercivity constant appearing in (4.5).

Remark 9.1. The constant k_4, unlike $k_1, k_2, k_3, k_5, \ldots$, depends only on the nonlinear operator $N(u)$ and not on the size of the initial data for equation (2.1).

Let us check (I) and (II). Let $u \in \Gamma$ be fixed and denote $\Lambda'_1 \leq \Lambda'_2 \leq \cdots \leq \Lambda'_n$ the eigenvalues of $P(u)AP(u)$. Thus $\Lambda(u) = \Lambda'_n$. We know that $0 \leq \Lambda(u) - \lambda_n$. On the other hand, $\Lambda'_1 + \cdots + \Lambda'_{n+1} \geq \lambda_1 + \cdots + \lambda_{n-1}$ (see [DS]). Therefore

$$0 \leq \Lambda(u) - \lambda_n \leq \Lambda'_n + \cdots + \Lambda'_n - (\lambda_1 + \cdots + \lambda_n). \tag{9.5}$$

Similarly, for $\lambda(u)$ we can find a vector such that $|h| = 1$, $P(u)h = 0$, $(I - P(u))Ah = \lambda(u)h$. We consider the linear space $M = \mathbb{R}h + P(u)H$. If P_M is the orthogonal projector on M then $\operatorname{Tr} P_M A P_M = \Lambda'_1 + \cdots + \Lambda'_n + \lambda(u)$ and therefore $\Lambda'_1 + \cdots + \Lambda'_n + \lambda(u) \geq \lambda_1 + \cdots + \lambda_n + \lambda_{n+1}$. Since $0 \leq \lambda_{n+1} - \lambda(u)$, we infer that

$$0 \leq \lambda_{n+1} - \lambda(u) \leq \Lambda'_1 + \cdots + \Lambda'_n - (\lambda_1 + \cdots + \lambda_n). \tag{9.6}$$

From (9.5) and (9.6) we conclude that (I) and (II) will be satisfied if

$$\Lambda_1' + \cdots + \Lambda_n' - (\lambda_1 + \cdots + \lambda_n) < \frac{\lambda_{n+1} - \lambda_n}{2}, \quad \text{that is, if}$$

$$\operatorname{Tr} AP(u) - \operatorname{Tr} AP_n < \frac{\lambda_{n+1} - \lambda_n}{2}. \tag{9.7}$$

Let us denote P_r the projector on $T_u(\Gamma)$. Then

$$P_n = \frac{u \otimes u}{|u|^2} + P_r \tag{9.8}$$

since $u/|u|$ is orthogonal on $T_u(\Gamma)$. (The notation $(a \otimes a)/|a|^2$ was introduced in Chapter 4.) Also

$$P(u) = \frac{w_0 \otimes w_0}{|w_0|^2} + P_r$$

where

$$w_0 = (I - P_r)N(u).$$

Thus

$$\operatorname{Tr} AP(u) - \operatorname{Tr} AP_N = \frac{|A^{1/2} w_0|^2}{|w_0|^2} - \frac{|A^{1/2} u|^2}{|u|^2}.$$

From (9.8) we deduce

$$(I - P_r) = I - P_n + \frac{u \otimes u}{|u|^2}$$

and therefore

$$w_0 = (I - P_r)N(u) = (I - P_n)N(u) + \frac{(N(u), u)}{|u|^2} u.$$

Now $u \in P_n H$ and the decomposition of w_0 gives

$$|A^{1/2} w_0|^2 = |(I - P_n)A^{1/2} N(u)|^2 + \frac{(N(u), u)^2}{|u|^4} |A^{1/2} u|^2,$$

$$|w_0|^2 = |(I - P_n)N(u)|^2 + \frac{(N(u), u)^2}{|u|^2}.$$

We obtain

$$\operatorname{Tr} AP(u) - \operatorname{Tr} AP_n = \frac{|u|^2 |A^{1/2}(I - P_n)N(u)|^2 - |A^{1/2} u|^2 |(I - P_n)N(u)|^2}{|u|^2 |(I - P_n)N(u)|^2 + (N(u), u)^2}. \tag{9.9}$$

Thus, (9.7) will be satisfied if

$$\frac{|u|^2|A^{1/2}(I - P_n)N(u)|^2}{(N(u), u)^2} < \frac{\lambda_{n+1} - \lambda_n}{2}. \tag{9.10}$$

We can already use $(N(u), u) \geq (k_4/2)[|A^{1/2}u|^2 + |u|^2]$ if $|u| = R$ with R chosen in (9.4). Then (9.10) follows from

$$\frac{|u|^2|A^{1/2}(I - P_n)N(u)|^2}{[|A^{1/2}u|^2 + |u|^2]^2} \leq \frac{k_4^2}{8}(\lambda_{n+1} - \lambda_n). \tag{9.11}$$

Now $N(u) = Au + B(u, u) + Cu + f$ and $(I - P_n)Au = A(I - P_n)u = 0$; therefore $(I - P_n)N(u) = (I - P_n)[B(u, u) + Cu + f]$ contains only "lower order" terms and (9.11) can be realized provided n is large enough. We shall make this statement more precise in the last chapters, in which we apply our results to specific examples. Let us now state the information gathered in this section as Proposition 9.2.

Proposition 9.2. *Let $\gamma > 0$ be fixed. Let R satisfy $R \geq (2/k_4)\sqrt{|A^{1/2}\varphi|^2 + |\psi|^2}$ (see 9.4)) and $\frac{1}{2}R \geq \rho_0 + 4\lambda_{n+1}^{-1/4}\rho_1$ (see (6.4)). Let n satisfy*

$$\frac{|u|^2|A^{1/2}(I - P_n)N(u)|^2}{[|A^{1/2}u|^2 + |u|^2]^2} \leq \frac{k_4^2}{8}(\lambda_{n+1} - \lambda_n) \quad (\text{see } (9.1))$$

and $\lambda_{n+1} \geq (1/2\gamma^2)(\lambda_{n+1} - \lambda_n)$. Then the $(n-1)$-dimensional sphere

$$\Gamma = \{u | P_n u = u, |u| = R\}$$

satisfies

(I) $\Lambda(u) < \frac{1}{2}(\lambda_n + \lambda_{n+1})$, $u \in \Gamma$,
(II) $\lambda(u) > \frac{1}{2}(\lambda_n + \lambda_{n+1})$, $u \in \Gamma$,
(III) $(N(u), u) \geq \frac{k_2}{2}[|u|^2 + |A^{1/2}u|^2]$, $u \in \Gamma$,
(IV) $\Gamma \subset C_{n,X} = \bigcap_{x \in X}\{u | |(I - P_n)(u - x)| \leq \frac{1}{3}|P_n(u - x)|\}$,
(V) *For every $u \in \Gamma$, every $w \in N(u)\mathbb{R} + T_u(\Gamma)$ satisfies $|(I - P_n)w| \leq \gamma|P_n w|$.*

Proof. The only properties that were not checked were (IV) and (V). Now (IV) follows from Corollary 6.4 and (V) follows from the fact that $|(I - P_n)w| \leq \gamma|P_n w|$ is a consequence of $|(I - P_n)N(u)| \leq \gamma(N(u), u)/|u|$, which in turn is a consequence of (9.10) and $\lambda_{n+1} \geq (1/2\gamma^2)(\lambda_{n+1} - \lambda_n)$. □

Remark 9.3. If condition (9.1) is not fulfilled or if the nature of the nonlinearities is such that (9.11) cannot be achieved on a sphere, one is led to consider more general surfaces $\Gamma \subset P_n H$. Suppose Γ is the smooth oriented boundary of the bounded, connected open set D. Properties (I) and (II) will hold if

$$\frac{|A^{1/2}(I - P_n)N(u)|^2}{(N(u), \nu(u))^2} < \frac{\lambda_{n+1} - \lambda_n}{2} \tag{9.12}$$

where $v(u)$ is the unit outward normal to Γ (see (9.10)). Property (III) is replaced by $(N(u), v(u)) > 0$. Property (V) holds if $|(I - P_n)N(u)| \leq \gamma|(N(u), v(u))|$ and thus if (9.12) and $\lambda_{n+1} \geq (1/2\gamma^2)(\lambda_{n+1} - \lambda_n)$ are true.

Remark 9.4. We emphasize the fact that the results in this chapter are of purely geometric nature. They do not involve the dynamical system $(S(t))_{t \geq 0}$.

CHAPTER 10

Construction of the Inertial Manifold

Let Γ be the $(n-1)$-dimensional sphere $\{u | u = P_n u, |u| = R\}$ in $P_n H$ considered in Proposition 9.2. Let Σ be the integral manifold obtained with Γ as initial data:

$$\Sigma = \bigcup_{t>0} S(t)\Gamma. \tag{10.1}$$

Notice that at this state we do not yet know that Σ is a genuine manifold, but under appropriate conditions, to be specified below, Σ is even a smooth manifold. Actually, our aim is to show even more, namely that $\bar{\Sigma}$, the closure of Σ in H, is an inertial manifold for $(S(t))_{t\geq 0}$. Most of our arguments will have a geometric nature and therefore it is easier to follow them by drawing appropriate figures.

We start by listing the conditions under which our arguments work.

(C1) n, γ, R satisfy the conditions in Proposition 9.2 with $\gamma = \frac{1}{3}$;
(C2) $\Gamma \subset \theta Y$;
(C3) n satisfies (5.6); thus in particular $\lambda_{n+1} = \Lambda_{m'+1}, \lambda_n = \Lambda_{m'}$ for some m';
(C4) $m = m'$ satisfies (3.13);
(C5) $\lambda_1 + \cdots + \lambda_m > (k_9^2 + 2k_9')m$ for all $m \geq n - 2$.

We shall check the consistency for all large values of n in the case of the Kuramoto–Sivashinsky equations (see Chapter 15). For other equations (see Chapters 16 to 19) these conditions become simpler, although Γ is no longer a sphere in $P_n H$.

From now on we assume that the above conditions hold and we start the study of Γ by considering the application $s(t, u) = S(t)u$ for $t > 0$ and $u \in \Gamma$. This is a smooth function from $R_+ \times \Gamma$ into H. Let us also introduce $\sigma = P_n s$: $\mathbb{R}_+ \times \Gamma \to P_n H$. The Jacobian of σ at $[t_0, u_0]$ is given by

$$D\sigma(t_0, u_0) = [-P_n N(s(t_0, u_0)), P_n s'(t_0, u_0)],$$

where $N(u) = Au + R(u)$ and $s'(t_0, u_0)$ is the application that assigns to v_0 the value $v(t_0)$ of the solution of (2.3), the linearized equation along $S(t)u_0$.

Let us assume that for some $(t_0, u_0) \in R_+ \times \Gamma$, $(D\sigma)(t_0, u_0)$ is not invertible. Then there exists a tangent vector to Γ, $v_0 \in P_nH$, and a real number ξ such that the vector $w(t) = \xi N(S(t, u_0)) + s'(t, u_0)v_0$ satisfies $P_n w(t_0) = 0$, $w(t_0) \neq 0$. Now let $P(t)$ be the projector on the linear space $N(S(t, u_0))\mathbb{R} + s'(t, u_0)(T_{u_0}(\Gamma)$ denotes the tangent hyperplane (in P_nH) to Γ at u_0. Observe that $P(t)w(t) = w(t)$. Also $P(t)$ solves the transport equation (2.11) with initial data $P(0) = P(u_0)$ where $P(u_0)$ is the projector on $N(u_0)\mathbb{R} + T_{u_0}(\Gamma)$. The choice of Γ implies that $\Lambda(u_0) < (\lambda_n + \lambda_{n+1})/2$. From Corollary 3.1 (spectral blocking) we infer that $\Lambda(t_0) = \max\{(Ag, g) | P(t_0)g = g, |g| = 1\}$ satisfies $\Lambda(t_0) < (\lambda_n + \lambda_{n+1})/2$. Therefore $(Aw(t_0), w(t_0)) < (\lambda_n + \lambda_{n+1})/2|w(t_0)|^2$. But since $P_n w(t_0) = 0$, $(Aw(t_0), w(t_0)) \geq \lambda_{n+1}|w(t_0)|^2$, and since $w(t_0) \neq 0$ we obtain the contradiction $\lambda_{n+1} < \lambda_n$.

We thus proved the first step in the construction of the inertial manifold:

Step 1. $\sigma = P_n s(t, u)$ is regular at each $(t, u) \in (0, \infty) \times \Gamma$.

In particular, we infer that σ is locally invertible. If $u = s(t_0, u_0)$ is a point on Σ and if σ_0^{-1} is the inverse of σ on a neighborhood U_0 of $\sigma(t_0, u_0) = P_n u$, then the collection (u_0, Ω_0), $\Omega_0 = s \cdot \sigma_0^{-1} : U_0 \to \Sigma \subset H$ forms an atlas for Σ. Thus $P_n\Sigma$ is open in P_nH. Now Γ was chosen to satisfy $\Gamma \subset C_{n,X} \cap \theta Y$ where

$$C_{n,X} = \bigcap_{x \in X} \{u | |(I - P_n)(u - x)| \leq \tfrac{1}{3}|P_n(u - x)\}$$

(see condition (IV) of Proposition 9.1). We saw (Chapter 5) that the strong squeezing property implies that $S(t)(C_{n,X} \cap \theta Y) \subset C_{n,X}$. Therefore $\Sigma \subset C_{n,X}$. Now clearly $\Sigma \cap X$ is void. For suppose $u \in \Sigma \cap X$. Then $u = S(t_0, u_0)$ for some $u_0 \in \Gamma$, $t_0 > 0$. But $S(t_0)X = X$ and therefore by the injectivity of $S(t_0)$, $u_0 \in \Gamma \cap X$. But $\Gamma \cap X = \varnothing$ since we required $|u_0| = R > \rho_0$. The fact that $\Sigma \cap X = \varnothing$ together with $\Sigma \subset C_{n,X}$ implies obviously $P_nX \cap P_n\Sigma = \varnothing$.

Now let $p \in \overline{P_n\Sigma} \backslash P_n\Sigma$. There exists a sequence $(t_n, u_n) \in R_+ \times \Gamma$ such that $P_n s(t_n, u_n)$ converges to p and u_n converges to some $u_0 \in \Gamma$. The sequence $(t_n)_{n=1}^{\infty}$ cannot have a cluster point $t_0 \in (0, \infty)$. Indeed, if it had, one would have found a subsequence $t_{n_k} \to t_0$ and thus $p = \lim_{k\to\infty} P_n(s(t_{n_k}, u_{n_k}) = P_n s(t_0, u_0) \in P_n\Sigma$ is absurd. If 0 is a cluster point for $\{t_n\}_{n=1}^{\infty}$ then p must belong to Γ. If there exists $t_{n_k} \to \infty$ then we claim p must belong to the projection of the universal attractor X. Indeed, we may assume that $s(t_{n_k}, u_{n_k})$ converges to some u_∞. Clearly, $P_n u_\infty = p$. On the other hand, $s(t_{n_k}, u_{n_k}) = S(t)s(t_{n_k} - t, u_{n_k})$ for any $t > 0$ and $t_{n_k} > t$. We conclude from the fact that $s(t_{n_k} - t, u_{n_k})$ are all contained in some large ball B in H that $u_\infty \in S(t)B$, for all $t > 0$, and therefore $u_\infty \in X$. We proved therefore:

Step 2. $P_n\Sigma$ is an open set in H, $P_n\Sigma \cap P_nX = \varnothing$ and $\overline{P_n\Sigma} \subset P_nX \cup P_n\Sigma \cup \Gamma$.

The next step is to prove that

Step 3. $\overline{P_n\Sigma} \supset \{P_n u \,|\, |u| \le R\}$.

Since $\Gamma = P_n\Gamma \subset \overline{P_n\Sigma}$, it is enough to prove that $\overline{P_n\Sigma}$ covers the open ball in P_nH of radius R. Let us reason by contradiction and assume that there exists an $\varepsilon > 0$ such that the ball $B_n = \{p \,|\, P_n p = p, |p - p_0| \le \varepsilon\}$ for some $p_0 \in P_nH$ is included in $\{|p| < R\}$ but does not intersect $P_n\Sigma$. It follows that $B_n \cap P_n s(t, \Gamma) = \varnothing$ for any $t > 0$. From the isoperimetric inequality it follows that

$$\mathrm{vol}_{n-1}(\partial B_n) \le \mathrm{vol}_{n-1}[P_n s(t, \Gamma)] \le \mathrm{vol}_{n-1} s(t, \Gamma), \qquad t > 0.$$

But $\mathrm{vol}_{n-1} s(t, \Gamma)$ decreases exponentially by condition (C5) and Corollary 8.2, absurd.

Now it is obvious that Σ is connected (even arcwise connected); therefore $P_n\Sigma$ is connected. By virtue of Steps 1 and 2, $(\Sigma, P_n|\Sigma)$ is a covering space of the open connected set $P_n\Sigma$. Therefore the cardinal of $p_n^{-1} p \cap \Sigma$ is equal to the number of connected exponents of Σ, that is, equal to one. Thus we established

Step 4. The map $P_n : \Sigma \to P_nH$ is injective. Moreover, $\overline{P_n\Sigma} = P_n\Sigma \cup P_nX \cup \Gamma \supset \{P_n u = u, |u| \le R\}$ and $P_n\Sigma \cup P_nX \supset \{u \,|\, |u| < R, P_n u = u\}$. We therefore define $\Phi : \{u \,|\, |u| \le R, P_n u = u\} \to H$ by

$$\Phi(p) = \begin{cases} p & \text{if } |p| = R, \\ u \in \Sigma \text{ satisfying } P_n u = p & \text{if } p \in P_n\Sigma, |p| < R, \\ u \in X \text{ satisfying } P_n u = p & \text{if } p \in P_nX. \end{cases} \tag{10.2}$$

The next step is to show that Φ is Lipschitz. Let p_1, p_2 be two points in P_nH, $|p_1| \le R$, $|p_2| \le R$. Let us consider the straight line that joins them: $p(\tau) = p_1 + \tau(p_2 - p_1)$, $\tau \in [0, 1]$. Assume first that $p(\tau) \notin X$ for all $\tau \in [0, 1]$. There exists then a smooth curve $u(\tau)$ in Σ such that $p(\tau) = P_n u(\tau)$, $\tau \in (0, 1)$. Let us take $\tau \in (0, 1)$ and consider $T_{u(\tau)}\Sigma$ the tangent linear manifold to Σ. Since $u(\tau) = s(t_0, u_0)$ for some $t_0 > 0$ and $u_0 \in \Gamma$ it follows that $du/d\tau \in T_{u(\tau)}(\Sigma)$ is the value $w(t_0)$ of the solution of the linearized equation (2.3) along $S(t)u_0$ with initial data $w(0)$ belonging to $N(u_0)\mathbb{R} + T_{u_0}(\Gamma)$. From Theorem 5.1 and (V) of Proposition 9.2 it follows that

$$\left|(I - P_n)\frac{du}{ds}\right| \le \frac{1}{3}\left|P_n\left(\frac{du}{ds}\right)\right|.$$

Therefore

$$\begin{aligned} |(I - P_n)(u(0) - u(1))| &\le \int_0^1 \left|(I - P_n)\frac{du}{d\tau}\right| d\tau \\ &\le \frac{1}{3}\int_0^1 \left|P_n \frac{du}{d\tau}\right| d\tau = \frac{1}{3}\int_0^1 \left|\frac{d}{d\tau}(p_1 + \tau(p_2 - p_1))\right| d\tau \\ &= \frac{1}{3}|p_2 - p_1|. \end{aligned}$$

Therefore, we proved that, as long as $p(\tau)$ does not encounter P_nX,

$$|\Phi(p_1) - \Phi(p_2)| \leq \tfrac{4}{3}|p_1 - p_2|$$

and

$$|(I - P_n)[\Phi(p_1) - \Phi(p_2)]| \leq \tfrac{1}{3}|p_1 - p_2|.$$

Suppose now that there exists $\tau \in (0, 1)$ such that $p(\tau) \in P_nX$. Let u_∞ be such that $P_n u_\infty = p(\tau)$. Since $\Sigma \cup X \subset C_{n,X}$ we have that, if $u_i = \Phi(p_i)$, $i = 1, 2$,

$$|(I - P_n)(u_i - u_\infty)| \leq \tfrac{1}{3}|P_n(u_i - u_\infty)|, \qquad i = 1, 2.$$

Now

$$\begin{aligned} |(I - P_n)(u_1 - u_2)| &\leq |(I - P_n)(u_1 - u_\infty)| + |(I - P_n)(u_2 - u_\infty)| \\ &\leq \tfrac{1}{3}[|p_1 - p(\tau)| + |p_2 - p(\tau)|] = \tfrac{1}{3}|p_1 - p_2|. \end{aligned}$$

We thus proved Step 5.

Step 5. $|(I - P_n)[\Phi(p_1) - \Phi(p_2)]| \leq \frac{1}{3}|p_1 - p_2|$, $p_1, p_2 \in P_nH$, $|p_1|, |p_2| \leq R$.

It is noteworthy to remark that from Step 5 and the coercivity property (III) (in Chapter 9) it follows that

$$P_n\Sigma = \{p \mid p \in P_nH, p \notin P_nX, |p| < R\}. \tag{10.3}$$

The last step will treat the problem of uniform exponential convergence toward $\bar{\Sigma}$ of any solution $S(t)u_0$. Therefore let u_0 be arbitrary in $H \backslash X$. Since all trajectories enter in finite time in Y we may assume $u_0 \in Y$. We recall that Theorem 6.2 states that either

$$\operatorname{dist}(S(t)u_0, X) \leq \max_{x \in X} |u_0 - X| \exp(-k_8 t) \quad \text{for all } t > 0$$

(and in this case we have proved the exponential convergence of $S(t)u_0$ toward $\bar{\Sigma}$), or there exists $t_0 > 0$ such that for $t \geq t_0$, $S(t)u_0$ belongs to $C_{n,X}$. In this last case, let $u = S(t)u_0 \in C_{n,\underline{X}}$ for some $t \geq t_0$; we can also assume that u is so close to X that $p = P_nu \in P_n\bar{\Sigma}$. It follows that $p \in P_n\Sigma$. Indeed, the only other possibility, $p \in P_nX$, is excluded by $u \in C_{n,X}$. Therefore there exists $u_1 \in \Sigma$ such that $p = P_nu_1$. Let $x \in X$ be arbitrary. We claim that $|u - u_1| < |u - x|$. Indeed, since $u \in C_{n,X}$, $|(I - P_n)(u - x)| \leq \frac{1}{3}|P_n(u - x)|$, and since $\Sigma \subset C_{n,X}$, $|(I - P_n)(u_1 - x)| \leq \frac{1}{3}|P_n(u_1 - x)|$. But $u - u_1 = (I - P_n)(u - u_1)$ and $P_n(u - x) = P_n(u_1 - x)$; therefore $|u - u_1| \leq \frac{2}{3}|P_n(u - x)| \leq \frac{2}{3}|u - x|$, since $u \notin X$, $|u - x| > 0$, and thus $\frac{2}{3}|u - x| < |u - x|$. We can also assume that $t \geq t_0$, $S(t)u$ is nearer to X than to Γ. (Notice that because of the definitions of X, Y, and Γ, we can choose this t_0 depending only on $|u_0|$.) It follows that the distance from u to $\bar{\Sigma}$ is attained on Σ and by virtue of conditions (C1) and (C4) we can apply the results in Chapter 7 and deduce that

$$\operatorname{dist}[S(t)u_0, \bar{\Sigma}] \leq \operatorname{dist}[S(t_0)u_0, \bar{\Sigma}] \exp(-k_8 t), \qquad t \geq t_0.$$

We thus proved Step 6.

Step 6. For any $u_0 \in H$, there exists $t_0 \geq 0$ (t_0 depending on $|u_0|$ only) such that

$$\operatorname{dist}(S(t)u_0, \bar{\Sigma}) \leq k_{11} \exp(-k_8 t), \qquad t \geq t_0.$$

(Here k_{11} depends only on R.)

Let us now consider $u \in \Sigma$ and denote by $P(u)$ the orthogonal projector in H on the tangent linear manifold to Σ at u. Denoting $\Lambda(u) = \Lambda(P(u))$ and $\lambda(u) = \lambda(P(u))$ we observe that (C4) and (C1) together with the spectral blocking property (Corollary 3.3) imply that

$$\Lambda(u) < \frac{\lambda_n + \lambda_{n+1}}{2} \quad \text{and} \quad \lambda(u) > \frac{\lambda_n + \lambda_{n+1}}{2}.$$

Also we notice that for $p \in P_n H$, $0 < |p| \leq R$, we have

$$|\Phi(p)|^2 = |p|^2 + \left|(I - P_n)\left(\Phi(p) - \Phi\left[\frac{R}{|p|}p\right]\right)\right|^2 \leq |p|^2 + \tfrac{1}{9}(R - |p|)^2 \leq R^2$$

whence $\bar{\Sigma} \subset \{u \in H \mid |u| \leq R\}$.

Summarizing, we proved Theorem 10.1.

Theorem 10.1. *Let* $\Gamma = \{u \mid P_n u = u, |u| = R\}$ *and let all conditions* (C1) *to* (C5) *hold. Then the closure* $\bar{\Sigma}$ *of the smooth integral manifold* $\Sigma = \bigcup_{t>0} S(t)\Gamma$ *is an inertial manifold in the ball* $\{u \in P_n H \mid |u| \leq R\}$. *Moreover, precisely:*

(a) *The projection* P_n *restricted to* $\bar{\Sigma}$ *is injective and its inverse* $\Phi : \{u \mid P_n u = u, |u| \leq R\} \to \bar{\Sigma} \subset H$ *satisfies*

$$|(I - P_n)[\Phi(p_1) - \Phi(p_2)]| \leq \tfrac{1}{3}|P_n[\Phi(p_1) - \Phi(p_2)]|$$

for all p_1, p_2 *in* $P_n H$, $|p_1| \leq R$, $|p_2| \leq R$.

(b) *For every* $u \in \Sigma$ *the projector* $P(u)$ *on the tangent space at* u *to* Σ *satisfies*

$$\Lambda(P(u)) < \frac{\lambda_n + \lambda_{n+1}}{2}, \qquad \lambda(P(u)) > \frac{\lambda_n + \lambda_{n+1}}{2}.$$

(c) *For any* $u_0 \in H$, *there exists* $t_0 \geq 0$ *depending on* $|u_0|$ *only, such that for* $t \geq t_0$

$$\operatorname{dist}(S(t)u_0, \bar{\Sigma}) \leq k_{11} \exp(-k_8 t), \qquad t \geq t_0,$$

with a constant k_{11} *depending only on* R.

Remark 10.2. If Γ is the smooth oriented boundary of the bounded open connected set $D \subset P_n H$ and if Γ satisfies (I) to (V) (see Remark 9.3) then Theorem 10.1 is valid, provided some appropriately modified conditions (C1) to (C5) hold. Φ will be defined $\Phi : \bar{D} \to \bar{\Sigma}$. (See Chapters 17 and 18.)

CHAPTER 11

Lower Bound for the Exponential Rate of Convergence to the Attractor

Let $\bar{\Sigma}$ be the inertial manifold constructed in Chapter 10. We recall that Σ is smooth and that $\bar{\Sigma}$ is parametrized by $B_n = \{u | P_n u = u, |u| \leq R\}$ through the Lipschitz function $\Phi : \bar{B}_n \to \bar{\Sigma}$ of the Lipschitz constant $\frac{4}{3}$. Thus, as long as $p = P_n p, |p| < R, \Phi(p) \in \Sigma$, it follows that $\partial\Phi(p)/\partial p_i, i = 1, \ldots, n$, satisfy $|\partial\Phi(p)/\partial p_i| \leq \frac{4}{3}$. Let B denote the ball in H, $B = \{u | |u| < R\}$.

Lemma 11.1. *Let $u \in X$ and let $\delta > 0$ be a constant. Then*

$$\operatorname{vol}_n(\bar{B}_\delta(u) \cap \Sigma) \leq (\tfrac{4}{3})^n \delta^n \omega_n \tag{11.1}$$

where $\bar{B}_\delta = \{u | |u - v| \leq \delta\}$ and ω_n is the volume of the unit ball in $\mathbb{R}^n$.

PROOF. Let $D = P_n(B_\delta(u) \cap \Sigma)$. Then

$$\operatorname{vol}_n(\overline{B_\delta(u)} \cap \Sigma) \leq \int_D \left| \frac{\partial \phi}{\partial p_1} \wedge \cdots \wedge \frac{\partial \phi}{\partial p_n} \right| dp \leq (\tfrac{4}{3})^n \operatorname{vol}_n D \leq (\tfrac{4}{3})^n \operatorname{vol}_n P_n B_\delta(P_n u).$$

□

Now let $\delta_X(t)$ denote the maximal distance between $u \in S(t)B$ and X:

$$\delta_X(t) = \sup_{u \in S(t)B} \operatorname{dist}(u, X). \tag{11.2}$$

Let us also denote by $n_\delta(X)$ the minimal number of balls centered in X of radii δ required to cover X. Let $\delta = \delta_X(t)$ for a fixed $t > 0$. Then by the definition of $n_\delta(X)$ there are $n_\delta(X)$ balls of radii δ that cover X, $X \subset \bigcup_{i=1}^{n_\delta(X)} B_\delta(u_i)$, $u_i \in X$. By the definition of $\delta_X(t)$ it follows that $S(t)B \subset \bigcup_{i=1}^{n_\delta(X)} B_{2\delta}(u_i)$. Therefore from (11.1) it follows that

$$\operatorname{vol}_n(S(t)\Sigma) \leq n_{\delta_X(t)}(X) [\tfrac{8}{3}]^n \delta_X(t)^n \omega_n. \tag{11.3}$$

Now $\delta_X(t) \to 0$ as $t \to \infty$. Thus, if $\varepsilon > 0$ there exists $t_0 = t_0(\varepsilon) > 0$ such that if $t \geq t_0$

$$\frac{\log n_{\delta_X(t)}(X)}{\log(1/\delta_X(t))} \leq d_M(X) + \varepsilon$$

where $d_M(X)$ is the fractal dimension of X. Therefore since $d_M(X) \leq n$ we obtain from (11.3)

$$\varliminf_{t\to\infty} \frac{\log \mathrm{vol}_n(S(t)\Sigma)}{t} + [n - d_M(X) - \varepsilon] \varlimsup_{t\to\infty} \frac{\log(1/\delta_X(t))}{t} \leq 0.$$

Since ε is arbitrary we proved

Proposition 11.2. *for Σ the n-dimensional integral manifold considered in Chapter 10 we have*

$$\varliminf_{t\to\infty} \frac{\log \mathrm{vol}_n(S(t)\Sigma)}{t} + (n - d_M(X)) \varlimsup_{t\to\infty} \frac{\log(1/\delta_X(t))}{t} \leq 0. \tag{11.4}$$

We shall give now a lower bound for

$$\varliminf_{t\to\infty} \frac{1}{t} \log \mathrm{vol}_n(S(t)\Sigma),$$

in case σ is the integral manifold considered in Chapter 10. Let us recall that we denoted by $s'(t, u_0)$ the first variation of $S(t)u$ along $S(t)u_0$. Let us consider $\psi: U$ (open $\subset$) $R^{n-1} \to \Gamma$ a parametrization at some point in Γ and denote $\Gamma_\psi = \Psi(U)$, $\Sigma_\psi = \bigcup_{t>0} S(t)\Gamma_\psi$. Then

$$\begin{aligned}
&\mathrm{vol}_n(S(t)\Sigma_\psi)\\
&= \int_t^\infty \int_U |s'(\tau, \psi)(\alpha))N(\psi(\alpha)) \wedge s'(\tau, \psi(\alpha))\partial_1\alpha \wedge \cdots \wedge s'(\tau, \psi(\alpha))\partial_{N-1}\alpha|\, d\alpha\, d\tau\\
&= \int_t^\infty \int_U \exp\left(-\int_0^\tau \mathrm{Tr}(A(\tau')P(\tau'))\, d\tau'\right)|N(\tau) \wedge \psi'_{\alpha_1}(\alpha) \wedge \cdots \wedge \psi'_{\alpha_{n-1}}(\alpha)|\, d\tau\, d\alpha.
\end{aligned} \tag{11.5}$$

Here $A(\tau')$ is the linearized $A + L(\tau')$ along $S(\tau')\psi(\alpha)$ and $P(\tau')$ is the projector on the space $s'(\tau', \psi(\alpha))N(\psi(\alpha))\mathbb{R} \to s'(\tau', \psi(\alpha))T_{\psi(\alpha)}(\Gamma)$. Now

$$\begin{aligned}
\mathrm{Tr}(A(t)P(t)) &\leq n \sup\{A(t)g, g) \mid |g| = 1, P(t)g = g\}\\
&\leq n \sup\{(Ag, g) + k_9|A^{1/2}g| + k'_9 | P(t)g = g, |g| = 1\}\\
&\leq n \sup\{2(Ag, g) + k_{12} | P(t)g = g, |g| = 1\}\\
&\leq n(2\Lambda(t) + k_{12}) \leq n(\lambda_n + \lambda_{n+1} + k_{12}).
\end{aligned}$$

We used (C3), $k_{12} = \frac{1}{4}k_9^2 + k'_9$, and the spectral blocking property (Corollary 3.3). It follows that

$$\mathrm{vol}_n(S(t)\Sigma_\psi) \geq \frac{\gamma_n}{n[\lambda_n + \lambda_{n+1} + k_{12}]} e^{-n[\lambda_n + \lambda_{n+1} + k_{12}]t} \tag{11.6}$$

with $\gamma_n = \int_U |N(\psi(\alpha)) \wedge \psi'_\alpha(\alpha) \wedge \cdots \wedge \psi'_{\alpha_{n-1}}(\alpha)|\, d\alpha$. Therefore

$$\underline{\lim}_{t\to\infty} \frac{1}{t} \log \operatorname{vol}_n S(t)\Sigma \geq -n[\lambda_n + \lambda_{n+1} + k_{12}]. \tag{11.7}$$

Combining (11.7) and (11.4) we obtain

Proposition 11.3. *Let Σ be the inertial manifold constructed in Theorem* 10.1. *Then*

$$\overline{\lim_{t\to\infty}} \log \frac{1}{\delta_X(t)} \leq \frac{\lambda_n + \lambda_{n+1} + k_{12}}{1 - d_M(X)/n}. \tag{11.8}$$

Obviously, (11.8) is a constraint on the dimension of the inertial manifold Σ. Indeed,

$$\lambda_0 = \overline{\lim_{t\to\infty}} \frac{1}{t} \log \frac{1}{\delta_X(X)} \quad \text{and} \quad d_M(X)$$

are independent of n, but n must satisfy

$$\left[1 - \frac{d_M(X)}{n}\right] \lambda_0 \leq \lambda_n + \lambda_{n+1} + k_{12}, \qquad n \geq d_M(X).$$

CHAPTER 12

Asymptotic Completeness: Preparation

Our aim here and the next two chapters is to prove two important properties of the inertial manifolds that are not (usually) satisfied by the attractors. In this chapter and Chapter 13 the property that we prove is *the asymptotic completeness* of the inertial manifold $\bar{\Sigma}$ that we have constructed. We recall that the asymptotic completeness means that given any orbit of the dynamical system, we can find another orbit lying on $\bar{\Sigma}$ that produces the same limit behavior at $t \to \infty$.

We will prove Theorem 12.1.

Theorem 12.1. *The assumptions are those of Theorem* 10.1, *and let* $\bar{\Sigma}$ *be the inertial manifold defined in that theorem. We also assume that*

$$\lambda_n + \lambda_{n+1} \geq k_4'' \quad (\text{see } (13.12)),$$

$$\frac{\lambda_{n+1} - \lambda_n}{2} \geq k_1'' + k_2'' \left[\frac{\lambda_n + \lambda_n}{2} \right]^{1/4} + k_3'' \left[\frac{\lambda_n + \lambda_{n+1}}{2} \right]^{1/2}$$

(*see* (13.13), (13.14), (13.22), (13.10a), (13.11a)) *where* k_1'', k_2'', k_3'', k_4'' *are some positive constants depending only on* θ. *Then for every* $u_0 \in H$, *there exists* $v_0 \in \bar{\Sigma}$ *and* $\alpha \geq 0$, *such that*

$$S(t)u_0 - S(t - \alpha)v_0 \to 0$$

as $t \to \infty$.

It is noteworthy that conditions (C3), (C4) and the second supplementary condition above are similar. In applications (as will be seen in Chapters 15 to 18) it suffices to check one of them with large enough constants.

This result will be proved in Chapter 13 after we present some preliminary facts (in Steps 1 to 3 below) that improve previous results and are interesting by themselves.

Step 1. $\bar{\Sigma}$ is Lipschitz for the $D(A^{1/2})$-norm.

We want to show that $\phi = (I - P_n)\Phi$ is Lipschitz in the $D(A^{1/2})$-norm, or more precisely from H into $\mathcal{D}(A^{1/2})$. Let $p_1 + \phi(p_1)$, $p_2 + \phi(p_2)$ be two points on $\bar{\Sigma}$, with $p_i \in P_n\bar{\Sigma}$. Then we have the following Lemma 12.2.

Lemma 12.2. *For every pair* $p_1, p_2 \in P_n\bar{\Sigma}$, *we have*

$$|A^{1/2}(\phi(p_1) - \phi(p_2))| \leq \left[\frac{\lambda_n + \lambda_{n+1}}{2}\right]^{1/2} \frac{1}{3} |p_1 - p_2|. \tag{12.1}$$

PROOF. The proof is different when the segment joining p_1 to p_2 (denoted $[p_1, p_2]$) meets $P_n X$ and when it does not. We first consider the case where $[p_1, p_2] \cap P_n X$ is empty. Then ϕ is continuously differentiable and not only Lipschitz on $[p_1, p_2]$ and we can write

$$\phi(p_2) - \phi(p_1) = \int_0^1 \phi'(p_1 + \tau(p_2 - p_1))(p_2 - p_1)\, d\tau,$$

thus

$$\begin{aligned} &|A^{1/2}(\phi(p_2) - \phi(p_1))| \\ &\leq \int_0^1 |A^{1/2}(\phi'(p_1 + \tau(p_2 - p_1))(p_2 - p_1))|\, d\tau \\ &\leq \text{(by the definition of } \Lambda \text{ in Chapter 3)} \\ &\leq \int_0^1 \Lambda(p_1 + \tau(p_2 - p_1))^{1/2} |\phi'(p_1 + \tau(p_2 - p_1))| |p_1 - p_2|\, d\tau. \end{aligned}$$

Now by Theorem 10.1(b) we have

$$\Lambda(p_1 + \tau(p_2 - p_1)) \leq \frac{\lambda_n + \lambda_{n+1}}{2}$$

since $p_1 + \tau(p_2 - p_1) \in P_n\Sigma$ for every τ, $0 \leq \tau \leq 1$. Also $|\phi'(p_1 + \tau(p_1 - p_1))| \leq \frac{1}{3}$ by Theorem 10.1(a); (12.1) follows in this case.

The second case is when $[p_1, p_2]$ intersects $P_n X$. We use condition (C5) (see Chapter 10) on n, which imposes that $n - 2$ is larger than or equal to the Hausdorff dimension of $P_n X$. We then consider in $P_n H$ two $(n - 1)$-dimensional balls D_1, D_2, centered at p_i, $i = 1, 2$, of sufficiently small radius, located in the $(n - 1)$-manifolds orthogonal to $[p_1, p_2]$ at p_1 and p_2; they limit a portion of the cylinder of axis $[p_1, p_2]$ based on D_1 (or D_2). Since the Lipschitz mappings do not increase the Hausdorff dimension, the projections of $P_n X$ on D_1 and D_2 parallel to $[p_1, p_2]$ have a Hausdorff dimension bounded by $n - 2$ and their $(n-1)$-Hausdorff measure is thus 0. This implies that we can find two sequences of points $\Pi_j^1 \in D_1$, $\Pi_j^2 \in D_2$, $j \in \mathbb{N}$, such that $\Pi_j^1 \to p_1$, $\Pi_j^2 \to p_2$ as $j \to \infty$, and the segment $[\Pi_j^1, \Pi_j^2]$ is parallel to $[p_1, p_2]$ and does not intersects $P_n X$. The first part of the proof shows that

$$|A^{1/2}(\phi(\Pi_j^1) - \phi(\Pi_j^2))| \leq \left[\frac{\lambda_n + \lambda_{n+1}}{2}\right]^{1/2} \frac{1}{3}|\Pi_j^1 - \Pi_j^2|, \qquad j = 1, 2, \ldots . \tag{12.2}$$

We will obtain (12.1) in this case by letting $j \to \infty$. This point is not, however, straightforward since we do not yet know that ϕ is continuous from $P_n H$ into $D(A^{1/2})$. (When proved, (12.1) will clearly imply this continuity property.) As $j \to \infty$, Π_j^α converges to p_α, strongly in H, $\alpha = 1, 2$, and since ϕ is continuous from $P_n H$ into H, $\phi(\pi_j^\alpha) \to \phi(p_\alpha)$ strongly in H too. Whence $\phi(\Pi_j^1) - \phi(\Pi_j^2)$ converges to $\phi(p_1) - \phi(p_2)$ strongly in H and, by (12.2), remains bounded in $D(A^{1/2})$. When $\phi(p_1) - \phi(p_2)$ belongs to $\mathscr{D}(A^{1/2})$ and $\phi(\Pi_j^1) - \phi(\Pi_j^2)$ converges to $\phi(p_1) - \phi(p_2)$ weakly in $D(A^{1/2})$ as $j \to \infty$. We finally obtain (12.1) in this case by passing to the lower limit in (12.2). □

Step 2. The rate of backward separation of trajectories on $\bar{\Sigma}$.

We know that two orbits can separate at an exponential rate backward or forward in time. Our aim here is to determine more precisely the rate at which two trajectories lying on $\bar{\Sigma}$ can separate backward in time.

We assume that $u_i = p_i + \phi(p_i)$, $i = 1, 2$, are two trajectories lying on Σ during an interval of time $-t_0 < t < 0$ $(t_0 > 0)$. We set $p = p_1 - p_2$, $w = u_1 - u_2$, $\tilde{p} = \frac{1}{2}(p_1 + p_2)$, $\tilde{u} = \frac{1}{2}(u_1 + u_2)$ and, for the sake of simplicity, we reverse time. We obtain the p satisfies the equation

$$-\dot{p} + Ap + P_n \tilde{L}(t) w = 0 \tag{12.3}$$

where $\tilde{L}(t)$ is the linearization of $R(u)$ around $\tilde{u}$ and properties (3.7), (3.8) are used. Thus

$$\begin{aligned} \frac{1}{2}\frac{d}{dt}|p|^2 &= |A^{1/2}p|^2 + (\tilde{L}(t)w, p) \\ &\leq |A^{1/2}p|^2 + |\tilde{L}(t)w||p| \leq (\text{by (3.6)}) \\ &\leq |A^{1/2}p|^2 + |p|[k_1^{1/2}|w| + k_2^{1/2}|A^{1/4}w| + k_3^{1/2}|A^{1/2}w|]. \end{aligned}$$

But

$$|w| = |p + \phi(p_1) - \phi(p_2)| \leq |p| + |\phi(p_1) - \phi(p_2)| \leq \tfrac{4}{3}|p|$$

and, by Lemma 12.2,

$$\begin{aligned} |A^{1/2}w| &\leq |A^{1/2}(p_1 - p_2)| + |A^{1/2}(\phi(p_1) - \phi(p_2))| \\ &\leq \left[\lambda_n^{1/2} + \frac{1}{3}\left(\frac{\lambda_n + \lambda_{n+1}}{2}\right)^{1/2}\right]|p| \leq \frac{4}{3}\left(\frac{\lambda_n + \lambda_{n+1}}{2}\right)^{1/2}|p|. \end{aligned}$$

Similarly,

$$|A^{1/4}w| \leq |w|^{1/2}|A^{1/2}w|^{1/2} \leq \frac{4}{3}\left(\frac{\lambda_n + \lambda_{n+1}}{2}\right)^{1/4}|p|.$$

Since $|A^{1/2}p| \leq \lambda_n^{1/2}|p|$, we now write

$$|p|\frac{d}{dt}|p| \leq \lambda_n|p|^2 + \frac{4}{3}\left[k_1^{1/2} + k_2^{1/2}\left(\frac{\lambda_n + \lambda_{n+1}}{2}\right)^{1/4} + k_3^{1/2}\left(\frac{\lambda_n + \lambda_{n+1}}{2}\right)^{1/4}\right]|p|^2,$$

that is,

$$\frac{d}{dt}|p| \leq \mu|p| \quad \text{where } \mu = \lambda_n + \frac{4}{3}\left[k_1^{1/2} + k_2^{1/2}\left(\frac{\lambda_n + \lambda_{n+1}}{2}\right)^{1/2} + k_3^{1/2}\left(\frac{\lambda_n + \lambda_{n+1}}{2}\right)^{1/2}\right]. \tag{12.4}$$

Thus

$$|p(t)| \leq |p(0)|e^{\mu t}$$

and since

$$|u|^2 = |p|^2 + |\phi(p_1) - \phi(p_2)|^2 \leq \tfrac{10}{9}|p|^2,$$

we obtain

$$|u| \leq \frac{\sqrt{10}}{3}|p(0)|e^{\mu t}.$$

Lemma 12.3. *If $u_i = p_i + \phi(p_i)$, $i = 1, 2$, are two trajectories lying on $\bar{\Sigma}$ for $-t_0 < t < 0$, then, for these values of t,*

$$\begin{cases} |p_1(t) - p_2(t)| \leq |p_1(0) - p_2(0)|e^{\mu|t|}, \\ |u_1(t) - u_2(t)| \leq |P_n(u_1(0) - u_2(0))|\dfrac{\sqrt{10}}{3}e^{\mu|t|} \end{cases} \tag{12.5}$$

where

$$\mu = \lambda_n + \frac{4}{3}\left[k_1^{1/2} + k_2^{1/2}\left(\frac{\lambda_n + \lambda_{n+1}}{2}\right)^{1/4} + k_3^{1/2}\left(\frac{\lambda_n + \lambda_{n+1}}{2}\right)^{1/2}\right]. \tag{12.6}$$

Step 3. The rate of convergence of the trajectories toward $\bar{\Sigma}$.

We know from Chapter 7 that the trajectories tend exponentially to $\bar{\Sigma}$ as $t \to \infty$. Our aim here is to slightly improve the rate at which the convergence takes place. We start by noticing that eventually any solution enters and stays in the set

$$\theta Y \cap \left\{ u \,|\, |u| \le \rho_0 + \frac{4\rho_1}{\lambda_{n+1}^{1/4}} \right\}$$

(see (6.3)) and that for u_1 in this set $\operatorname{dist}(u_1, \bar{\Sigma})$ is attained either in Σ or in X. We will prove Lemma 12.4.

Lemma 12.4. *Let $u = u(t)$ be a trajectory such that $u(t) \in \theta Y$, $|u(t)| \le \rho_0 + 4\rho_1/\lambda_{n+1}^{1/4}$, $t \ge 0$. Assume moreover that*

$$u(t) \in C'_{n,X} = \{ v \in H \,|\, v - x \in C_{n,1/2}, x \in X \} \quad \text{(see (5.4))}.$$

Then the distance $\operatorname{dist}(u(t), \bar{\Sigma})$ is bounded at $t \to \infty$ by $\operatorname{dist}(u(0), \bar{\Sigma}) \exp(-\sigma t)$, where

$$\sigma = \frac{\lambda_{n+1} + \lambda_n}{2} - k'_3 \left(\frac{\lambda_n + \lambda_{n+1}}{2} \right)^{1/2} - k'_1$$

(with $k'_1 = k_1^{1/2} + \frac{3}{4} k_2^{1/2}$, $k'_3 = k_3^{1/2} + \frac{1}{4} k_2^{1/2}$) provided $\lambda_n + \lambda_{n+1} > (k'_3)^2/2$.

Proof. Let $\delta(t) = \operatorname{dist}(u(t), \Sigma)$, $t \ge 0$. At some arbitrary time t_1, we consider $v_1 \in \Sigma \cup X$ such that $\delta(t_1) = |u(t_1) - v_1|$; we will compare for $t > t_1$ the orbits $u(t) = S(t - t_1)u(t_1)$ and $v(t) = S(t - t_1)v_1$, the second one lying on $\Sigma \cup X$. For $t > t_1$,

$$\delta(t) \le |u(t) - v(t)|.$$

Setting $w = u - v$, we have

$$\frac{1}{2} \frac{\delta^2(t) - \delta^2(t_1)}{t - t_1} \le \frac{1}{2} \frac{|w(t)|^2 - |w(t_1)|^2}{t - t_1}.$$

As $t \to t_1 + 0$, we see that the right upper derivative $\bar{D}_r \delta^2(t_1)$ at $t = t_1$ is bounded by $d|w(t)|^2/dt$ at $t = t_1$, and thus

$$\frac{1}{2} \bar{D}_r \delta^2(t_1) \le \frac{1}{2} \frac{d}{dt} |w(t_1)|^2.$$

We then repeat the computations in Chapter 4, using in particular (3.7) for $\tilde{L}(t)$. We obtain

$$\frac{1}{2} \frac{d}{dt} |w|^2 \le -|A^{1/2} w|^2 + k_1^{1/2} |w|^2 + k_2^{1/2} |w|^{3/2} |A^{1/2} w|^{1/2} + k_3^{1/2} |w| \, |A^{1/2} w|.$$

Now if $v_1 \in X$ then

$$\begin{aligned} |u(t_1) - [Pu(t_1) + \phi(Pu(t_1))]| &= |(I - P)u(t_1) - \phi(Pu(t_1))| \\ &\le |(I - P)(u(t_1) - v_1)| + |\phi(Pv_1) - \phi(Pu(t_1))| \\ &\le \tfrac{5}{6} |u(t_1) - v_1|, \end{aligned}$$

in contradiction with the definition of v_1. Thus $v_1 \in \Sigma$.

By property (C3) in Chapter 10 and the spectral blocking property (Corollary 3.3), and since $w(t_1)$ is orthogonal to the linear manifold tangent to Σ at v_1, we have

$$\xi^2 = \frac{|A^{1/2}w(t_1)|^2}{|w(t_1)|^2} \geq \lambda(v_1) \geq \frac{\lambda_n + \lambda_{n+1}}{2} \quad \text{(denoted } \mu^2 \text{ hereafter).}$$

whence

$$\tfrac{1}{2}\bar{D}_r\delta^2(t_1) \leq -|w|^2(\xi^2 - k_3^{1/2}\xi - k_2^{1/2}\xi^{1/2} - k_1^{1/2}) \leq -|w|^2(\xi^2 - k_3'\xi - k_1'),$$

with $k_1' = k_1^{1/2} + \frac{3}{4}k_2^{1/2}$, $k_3' = k_3^{1/2} + k_2^{1/2}/4$. But for n large, $\mu > k_3'/2$, and since $\xi \geq \mu$, we have

$$\xi^2 - k_3'\xi - k_1' \geq \mu^2 - k_3'\mu - k_1'.$$

Furthermore, this last quantity is positive if n is sufficiently large. Thus for n sufficiently large so that $\mu > k_3'/2$ we can write

$$\tfrac{1}{2}\bar{D}_2\delta^2(t_1) \leq -\delta^2(t_1)(\mu^2 - k_3'\mu - k_1'). \tag{12.7}$$

Actually, t_1 is arbitrary, and we obtain a rate of decay of δ,

$$\delta(t) \leq \delta(0)\exp(-(\mu^2 - k_3'\mu - k_1')t),$$

and the lemma follows. □

CHAPTER 13

Asymptotic Completeness: Proof of Theorem 12.1

In the previous chapter, Steps 1 to 3 do not depend on the validity of the asymptotic completeness property. In this chapter we assume that $u(t)$ never belongs to $\bar{\Sigma}$ since otherwise the result is obvious, and we argue by contradiction, assuming that Theorem 12.1 is not valid for $u(t)$. All further steps in this chapter will hinge on the negation of Theorem 12.1. Without loss of generality we can consider a trajectory $u(t) = S(t)u_0$ in $\theta Y \cap \{u \in H \mid |u| \leq R\}$. We assume that for every $v_0 \in \bar{\Sigma}$, $u(t) - v(t)$ (where $v(t) = S(t)v_0$) does not converge to 0 as $t \to \infty$. Thus for $v_0 \in \bar{\Sigma}$ fixed, there exists $\varepsilon > 0$ and a sequence $t_j \to \infty$ such that

$$|u(t_j) - v(t_j)| \geq \varepsilon > 0, \quad \text{for all } j. \tag{13.1}$$

Since $u(t)$ converges to $\bar{\Sigma}$ as $t \to \infty$, there exists t_* such that

$$\operatorname{dist}(u(t_j), \Sigma) < \eta\varepsilon \tag{13.2}$$

for $t_j \geq t_*$, where η, $0 < \eta < \frac{3}{7}$, is a fixed constant to be chosen later. Whence for such a j there exists $(p + \phi(p)) \in \Sigma$ for some $p = p_j$, such that $|u(t_j) - p + \phi(p))| \leq \eta\varepsilon$ and thus

$$\begin{aligned}
\varepsilon &\leq |v(t_j) - u(t_j)| \\
&\leq |v(t_j) - (p + \phi(p))| + |(p + \phi(p)) - u(t_j)| \\
&\leq \eta\varepsilon + |v(t_j) - (p + \phi(p))|.
\end{aligned}$$

We introduce the projectors $P = P_n$, $Q = I - P_n$, for a value of n satisfying the conditions in Theorem 10.1 and in Lemma 12.4. Then

$$\begin{aligned}
|v(t_j) - (p + \phi(p))| &\leq |Pv(t_j) - p| + |\phi(Pv(t_j)) - \phi(p)| \\
&\leq \text{(by the Lipschitz property of } \phi) \\
&\leq \tfrac{4}{3}|Pv(t_j) - p|.
\end{aligned}$$

Whence

$$\varepsilon \leq |v(t_j) - u(t_j)| \leq \eta\varepsilon + \tfrac{4}{3}|Pv(t_j) - p|. \tag{13.3}$$

By the definition of $p + \phi(p)$,

$$|Pu(t_j) - p| \leq |u(t_j) - (p + \phi(p))| \leq \eta\varepsilon$$

and

$$\begin{aligned} \varepsilon &\leq \eta\varepsilon + \tfrac{4}{3}|Pv(t_j) - Pu(t_j)| + \tfrac{4}{3}|Pu(t_j) - p| \\ &\leq \tfrac{7}{3}\eta\varepsilon + \tfrac{4}{3}|Pv(t_j) - Pu(t_j)|. \end{aligned}$$

We obtain

$$|Pv(t_j) - Pu(t_j)| \geq \frac{3 - 7\eta}{4}\varepsilon > 0, \quad \text{for all } t_j \geq t_*. \tag{13.4}$$

We then compare $Qv(t_j)$ and $Qu(t_j)$:

$$\begin{aligned} |Qv(t_j) - Qu(t_j)| &\leq |Qv(t_j) - \phi(p)| + |\phi(p) - Qu(t_j)| \\ &\leq \eta\varepsilon + |Qv(t_j) - \phi(p)| \leq \eta\varepsilon + |\phi(Pv(t_j)) - \phi(p)| \\ &\leq \eta\varepsilon + \tfrac{1}{3}|Pv(t_j) - p| \\ &\leq \eta\varepsilon + \tfrac{1}{3}|Pv(t_j) - Pu(t_j)| + \tfrac{1}{3}|Pu(t_j) - p| \\ &\leq \tfrac{4}{3}\eta\varepsilon + \tfrac{1}{3}|Pu(t_j) - Pv(t_j)|. \end{aligned}$$

Finally with (13.4)

$$|Qv(t_j) - Qu(t_j)| \leq \gamma|Pv(t_j) - Pu(t_j)|, \quad \text{for all } t_j \geq t_* \tag{13.5}$$

where choosing η small enough we have

$$\gamma = \frac{1}{3} + \frac{16\eta}{3(3 - 7\eta)} < \frac{1}{2}. \tag{13.6}$$

Since by conditions (C1) and (C3) (see Chapter 10), relation (5.6) is satisfied by $\gamma = \frac{1}{3}$, it will also be satisfied by the value of γ defined in (13.6) and by $\gamma = \frac{1}{2}$. By using Theorem 5.1 and relations (13.5), (13.6) we established Step 4.

Step 4. For every $v_0 \in \bar{\Sigma}$ there exists a time $t(v_0) \in [0, \infty)$ such that for $t < t(v_0)$, $w(t) = u(t) - S(t)v_0 \notin C_{n,\gamma}$ and $w(t) \in C_{n,\gamma}$ for $t \geq t(v_0)$.

Now $|Qw(t(v_0))| < \frac{1}{2}|Pw(t(v_0))|$ since $w(t(v_0)) \in C_{n,\gamma}$. (Since otherwise $w(t(v_0)) = 0$ and hence $u = u(t) \in \bar{\Sigma}$ for all $t \geq 0$, contradicting the choice of $u(\cdot)$.) Hence there exists a ball $B_{\rho_0}(v_0)$ centered at v_0 such that for all $v \in B_{\rho_0}(v_0) \cap \bar{\Sigma}$ the relation

$$|Q(u(t) - S(t)v)| < \tfrac{1}{2}|P(u(t) - S(t)v)| \tag{13.7}$$

holds for $t = t(v_0)$. Once again using Theorem 5.1 (with an appropriate γ for each $v \in B_{\rho_0}(v_0)$) we infer that (13.7) is valid for all $t \geq t(v_0)$, $v \in B_{\rho_0}(v_0) \cap \bar{\Sigma}$.

We can cover $\bar{\Sigma}$ by these balls $B_{\rho_0}(v_0)$, and since $\bar{\Sigma}$ is compact we can cover it by a finite number N of such balls: $B_{\rho_i}(v_i)$, $i = 1, \dots, N$. Then for every $v_* \in \bar{\Sigma}$, there exists i such that $v_* \in B_{\rho_i}(v_i)$ and (13.7) holds for every $t \geq t(v_i)$. Thus setting $t_* = \sup_{1 \leq i \leq N} t(v_i)$ we have proved Step 5.

Step 5. $S(t)v_0 - u(t) \in C_{n,1/2}$, for all $v_0 \in \bar{\Sigma}$, $t \geq t_*$.

Let $x \in X$ and $t \geq t_*$; there exists $v_0 \in X$ such that $x = S(t)v_0$ and therefore $x - u(t) \in C_{n,1/2}$. There follows Step 6.

Step 6. $u(t) \in C'_{n,X}$ for $t \geq t_*$.

Now let

$$\Sigma_2 = \bigcup_{t > t_*} S(t)\Gamma = S(t_*)\Sigma, \qquad \Sigma_1 = \bigcup_{0 \leq t \leq t_*} S(t)\Gamma. \tag{13.8}$$

It is clear that $\Sigma = \Sigma_1 \cup \Sigma_2$, $\bar{\Sigma} = \Sigma_1 \cup \bar{\Sigma}_2$, and Σ_2 contains the universal attractor X. Since the flow $-N(u)$ on Γ is pointing inside the manifold $S(t)\Gamma$, we notice that Σ_1 is limited by Γ and $S(t_*)\Gamma$ while Σ_2 is limited by $S(t_*)\Gamma$ and $\bar{\Sigma}_2$ constitutes an open neighborhood of X in $\bar{\Sigma}$. By Step 5,

$$|Q(u(t) - S(t)v_0)| \leq \tfrac{1}{2}|P(u(t) - S(t)v_0)|, \qquad t \geq t_*, v_0 \in \bar{\Sigma}.$$

Since v_0 is arbitrary in $\bar{\Sigma}$, $S(t)v_0$ can be an arbitrary point of $S(t)\bar{\Sigma}$. Whence

$$u(t) - v \in C_{n,1/2}, \quad \text{for all } t \geq t_*, v \in S(t)\bar{\Sigma}. \tag{13.9}$$

It follows from (13.9) that for $t \geq t_*$, $Pu(t)$ does not belong to the set $P\Sigma_2(t)$ where $\Sigma_2(t) = \bigcup_{\tau \geq t} S(\tau)\Gamma$, otherwise $Pu(t) = Pv$ for some $v \in S(\tau)\bar{\Sigma}$, $\tau > t$; then $v \in S(t)S(\tau - t)\bar{\Sigma} \subset S(t)\bar{\Sigma}$ and we can write (13.9) with $v = Pu(t) + \phi(Pu(t))$, i.e.,

$$|Qu(t) - \phi(Pu(t))| \leq \tfrac{1}{2}|Pu(t) - Pv| = 0.$$

Then $u(t) = v \in \bar{\Sigma}$, which is not possible as we assumed that $u(t)$ never belongs to $\bar{\Sigma}$. Since, for $t \geq t_*$, $Pu(t)$ does not belong to $P\Sigma_2(t)$, it must belong to $P\Sigma_1(t)$, where $\Sigma_1(t) = \bigcup_{0 < \tau < t} S(\tau)\Gamma$. Thus we established Step 7.

Step 7. For $t \geq t_*$, there exist $\tau = \tau(t) < t$ and $v = v(t) \in \Gamma$ such that

$$Pu(t) = PS(\tau(t))v(t).$$

The next step is to study the behavior of $\tau(t)$ as a function of t and the behavior of $\tau(t) - t$ as $t \to \infty$.

The norm of $u(t) - S(\tau(t))v(t)$ is equal to the vertical distance (i.e., in the QH direction) of $u(t)$ to $\bar{\Sigma}$, and an elementary computation shows that this is bounded by $(\sqrt{10}/3)\,\text{dist}(u(t), \bar{\Sigma})$. Thus

$$|u(t) - S(\tau(t))v(t)| \leq \frac{\sqrt{10}}{3}\,\text{dist}(u(t), \bar{\Sigma})$$

and by virtue of Step 6 and with Lemma 12.3

$$|u(t) - S(\tau(t))v(t)| \leq \left(2R\frac{\sqrt{10}}{3}\right)\exp(-\sigma t) \tag{13.10}$$

where σ is defined in Lemma 12.3:

$$\sigma = \frac{\lambda_{n+1} + \lambda_n}{2} - k_3'\left(\frac{\lambda_{n+1} + \lambda_n}{2}\right)^{1/2} - k_1'. \tag{13.10a}$$

According to Lemma 12.3, for any pair of points $v_1, v_2 \in \Gamma$ (or even $\bar{\Sigma}$), and for any $t \geq 0$,

$$\frac{\sqrt{10}}{3}|S(t)v_1 - S(t)v_2|\exp(\mu t) \geq |v_1 - v_2| \tag{13.11}$$

where μ is defined in Lemma 12.3:

$$\mu = \lambda_n + \frac{4}{3}\left(k_1^{1/2} + k_2^{1/2}\left(\frac{\lambda_n + \lambda_{n+1}}{2}\right)^{1/4} + k_3^{1/2}\left(\frac{\lambda_n + \lambda_{n+1}}{2}\right)^{1/2}\right). \tag{13.11a}$$

Recall that Lemma 12.4 holds under the supplementary condition

$$\lambda_n + \lambda_{n+1} > k_2 + k_3 \geq \frac{k_3'^2}{2}. \tag{13.12}$$

Now we impose the following much more stringent condition:

$$\sigma - \mu \geq k_{13} \tag{13.13}$$

where σ and μ are the parameters given by formulas (13.10a) and (13.11a), respectively, and k_{13} is a positive constant to be determined later.

The Lipschitz properties of the semigroup $(S(t))_{t\geq 0}$ are standard. They imply the existence of two constants k_{14}, k_{15} depending only on θ such that

$$|S(t)u - S(t)v| \leq k_{15}|u - v|\exp(k_{14}t) \tag{13.14}$$

for all $t \geq 0$, $u, v \in \theta Y$; in particular, for all $u = u(s)$, $s \geq 0$, and $v \in \bar{\Sigma}$. Now, due to (13.14), (13.10), we have for all $t \geq t_*$, $s \geq 0$,

$$\begin{aligned}
&|S(s + \tau(t))v(t) - S(\tau(s + t))v(s + t)| \\
&\quad\leq |S(s + \tau(t))v(t) - u(s + t)| + |S(\tau(s + t))v(s + t) - u(s + t)| \\
&\quad\leq |S(\tau(t))v(t) - u(t)|k_{15}\exp(k_{14}s) + \left(2R\frac{\sqrt{10}}{3}\right)\exp(-\sigma(s + t)) \\
&\quad\leq \left(2R\frac{\sqrt{10}}{3}\right)k_{15}\exp(k_{14}s - \sigma t) + \left(2R\frac{\sqrt{10}}{3}\right)\exp(-\sigma(t + s)).
\end{aligned}$$

By applying (13.11) we reached the next stage in our proof

Step 8. For $t \geq t_*$, $s > 0$, and

$$0 \leq T \leq T(s,t) = \min\{s + \tau(t), \tau(s + t)\} \tag{13.15}$$

we have

$$|S(s+\tau(t)-T)v(t)-S(\tau(s+t)-T)v(s+t)| \le k_{16}\exp(k_{14}s+\mu T-\sigma t) \tag{13.16}$$

where $k_{16}=(20(R/9))(k_{15}+1)$.

Our next aim is to show that $\delta(t)=\tau(t)-t$ is convergent for $t\to\infty$. This will be done in two steps.

First we take $T=T(s,t)$ in (3.16) and use either $\tau(t)\le t$ or $\tau(s+t)\le s+t$, obtaining

$$\begin{aligned}&|S(s+\tau(t)-T(s,t))v(t)-S(\tau(s+t)-T(s,t))v(t+s)|\\ &\quad\le k_{16}\exp[(k_{14}+\mu)s-k_{13}t].\end{aligned}$$

Denoting here by v either $v(t)$ or $v(s+t)$ and using $|v(t)|=|v(s+t)|=R$, we now deduce that

$$\begin{aligned}R-|S(|\delta(t)-\delta(t+s)|)v| &= R-|S(|s+\tau(t)-\tau(s+t)|)v|\\ &\le k_{16}\exp[(k_{14}+\mu)s-k_{13}t].\end{aligned} \tag{13.17}$$

But returning to our equation (2.1) and using (9.3) we have

$$\frac{d|u|^2}{dt}+k_4|u|^2\le 0 \quad \text{if } |u|\ge\frac{R}{2}. \tag{13.18}$$

(Here is the only time we use the assumption $R\ge(4/k_4)(|A^{1/2}\varphi|^2+|\psi|^2)^{1/2}$; otherwise we could have done with $R\ge(2/k_4)(|A^{1/2}\varphi|^2+|\psi|^2)^{1/2}$.) From (13.18) we infer

$$\frac{d|S(t)v|}{dt}+\frac{k_4}{4}R\le 0 \quad \text{if } |S(t)v|\ge\frac{R}{2}$$

whence

$$|S(t)v|\le R-\frac{k_4}{4}Rt \quad \text{if } |S(t)v|>\frac{R}{2},$$

which together with (3.17) yield

Step 9. For all $s\ge 0$, $t\ge t_*$ either

$$k_{17}\exp[(k_{14}+\mu)s-k_{13}t]>2 \tag{13.19}$$

or

$$|\delta(t)-\delta(t+s)|\le k_{18}\exp[(k_{14}+\mu)s-k_{13}t], \tag{13.20}$$

where $k_{17}=(\frac{80}{9})(k_{15}+1)$, $k_{18}=k_{17}/k_4$.

We can now prove that $\lim_{t\to\infty}\delta(t)$ exists and is $>-\infty$. Indeed, for $t\ge t_0\ge t_*$ with t_0 very large and $s=1$ we have only the alternative (13.20), and therefore

$$\sum_{m=0}^{\infty} |\delta(t_0 + m) - \delta(t_0 + m + 1)|$$
$$\le k_{17}[\exp(k_{14} + \mu)s - k_{13}t_0)] \sum_{m=0}^{\infty} (e^{-k_{13}})^m < \infty.$$

Thus

$$\lim_{m\to\infty} \delta(t_0 + m) = -t_\infty \quad \text{with } t_\infty \in [0, \infty). \tag{13.21}$$

Since for $t \to \infty$ we have $t = t_0 + m(t) + s(t)$ for some $s(t) \in [0, 1)$ and an integer $m(t) \to \infty$, we readily infer from (13.20) and (13.21) the following.

Step 10. $\lim_{t\to\infty} [\tau(t) - t] = \lim_{t\to\infty} \delta(t) = -t_\infty$ with $t_\infty \in [0, \infty)$.

Now let $\alpha > t_\infty$. Our last step before concluding the proof is the fact that for $t \to \infty$ the function $S(\alpha + \delta(t))v(t)$ converges fast enough. Notice that for t large enough, say $t \ge t_0 \ge t_*$, $\alpha + \delta(t) > 0$. Then in (13.16) we take $t \ge t_0$, $s > 0$, $T = t + s - \alpha$ and obtain

$$|S(\alpha + \delta(t))v(t) - S(\alpha + \delta(t + s))v(t + s)| \le k_{16} \exp[(k_{14} + \mu)s - \mu\alpha - k_{13}t). \tag{13.22}$$

Now we can use (13.22) exactly as we used (13.20) to establish Step 10. We thus obtain that $w = \lim_{t\to\infty} S(\alpha + \delta(t))v(t)$ exists; obviously $w \in \bar{\Sigma}$. Now for $t \ge t_0$

$$|w - S(\alpha + \delta(t))v(t)|$$
$$\le \sum_{m=0}^{\infty} |S(\alpha + (\delta(t + m + 1))v(t + m + 1) - S(\alpha + (\delta(t + m))v(t + m)$$
$$\le \sum_{m=0}^{\infty} k_{16} \exp[k_{14} + \mu - \mu\alpha - k_{13}(t + m)].$$

Summing this series we finally establish

Step 11. For $t \ge t_0$ $(\ge t_*)$ we have

$$|w - S(\alpha + \delta(t))v(t)| \le k_{16}(1 - e^{-k_{13}})^{-1}[\exp(k_{14} + \mu - \mu\alpha)] \exp(-k_{13}t). \tag{13.23}$$

We are now able to conclude the proof of Theorem 12.1. Indeed, from (13.10), (13.16) with $s = \alpha$ and $T = 0$, (13.23), and (13.14) we obtain for $t \ge t_0$

$$|u(t + \alpha) - S(t)w|$$
$$\le |u(t + \alpha) - S(\tau(t + \alpha))v(t + \alpha))| + |S(\tau(t + \alpha))v(t + \alpha) - S(\alpha + \tau(t))v(t)|$$
$$+ |S(\alpha + \tau(t))v(t) - S(t)w|$$

$$\leq \left(2R\frac{\sqrt{10}}{3}\right)\exp(-\sigma(t+\alpha)) + k_{16}\exp(k_{14}\alpha - \sigma t)$$

$$+ |S(t)S(\alpha + \delta(t))v(t) - S(t)w|$$

$$\leq M\exp(-\sigma t) + k_{15}k_{16}(1 - e^{-k_{13}})^{-1}\exp[k_{14} + \mu - \mu\alpha - (k_{13} - k_{14})t]$$

where M is independent of t. The theorem follows provided

$$k_{13} > k_{14} \tag{13.24}$$

CHAPTER 14

Stability with Respect to Perturbations

We prove in this chapter the stability of the inertial manifolds constructed before with respect to perturbations. Three types of perturbations will be explicitly considered here: perturbations of the operators corresponding to a Galerkin approximation of the problem, perturbation of the viscosity parameter v, and perturbation of the right-hand side f (see (14.1)). Although we restrict ourselves to these three perturbations for the sake of simplicity, we believe that our perturbation results apply to more general situations.

We consider a general perturbation of the basic equations (1.4), (2.1) of the form

$$\frac{d}{dt}u_\mu + \alpha_\mu A_\mu u_\mu + B_\mu(u_\mu, u_\mu) + C_\mu u_\mu + f_\mu, \tag{14.1}$$

$$u_\mu(0) = u_{0_\mu}, \tag{14.2}$$

where μ is an integer and, for each μ, A_μ, B_μ, C_μ are the Galerkin approximations of A, B, C:

$$A_\mu = P_\mu A, \qquad B_\mu = P_\mu B, \qquad C_\mu = P_\mu C, \qquad f_\mu = P_\mu f.$$

$f_\mu \in P_\mu H$, $u_{0_\mu} \in P_\mu H$, and $\alpha_\mu > 0$ is a viscosity coefficient that tends to 1 as $\mu \to \infty$.

Since A_μ, B_μ, C_μ, f_μ satisfy the same properties as A, B, C, f, it follows as indicated in Chapter 1 that for each μ, the problems (14.1), (14.2) possess a unique solution $u_\mu(t) = S_\mu(t)u_{0_\mu}$, where $S_\mu(t)$ is the perturbed nonlinear semigroup that converges to $S(t)$ (in the sense indicated below).

Also, for each μ, (14.1), (14.2) are similar to (2.1), (2.2) and the assumptions of Theorem 10.1 are satisfied for (14.1), (14.2) provided they are for (2.1), (2.1), which we shall suppose to be true in the sequel. Thus it is clear that Theorem

10.1 applies for every μ: it provides the existence of an integral manifold Σ_μ whose closure $\bar{\Sigma}_\mu$ is an inertial manifold of the perturbed system. (Recall that our aim is to show that $\bar{\Sigma}_\mu$ converges to $\bar{\Sigma}$ for an appropriate topology.) A careful examination of the proofs and discussions in the previous chapters shows that those results are valid uniformly with respect to μ, for $\mu \geq \mu_0$, when μ_0 is large enough. Thus if for n and R conditions (C1) to (C5) hold for (2.1), (2.2), they also hold for (14.1), (14.2), if $\mu \geq \mu_0$, $\mu \geq n$.

For each $\mu \geq \mu_0$, $\mu > n$, the inertial manifold $\bar{\Sigma}_\mu$ is given in the form

$$\Sigma_\mu = \{p + \phi_\mu(p), p \in D_n\}$$

where ϕ_μ is a Lipschitz function from $D_n = \{p \in P_n H \,|\, |p| \leq R\}$ into $(P_\mu - P_n)H$ and

$$|\phi_\mu(p_1) - \phi_\mu(p_2)| \leq \tfrac{1}{3}|p_1 - p_2| \quad \text{for all } p_1, p_2 \in D_n. \tag{14.3}$$

Also, the value of n necessary for the squeezing and invariance properties (Theorems 4.2 and 5.1) and the contraction of volumes of dimension $n - 1$ (Corollaries 8.2 and 8.3) is the same for all μ, $\mu \geq \mu_0$, and for the unperturbed system (conventionally corresponding to $\mu = \infty$). We now restrict ourselves to $\mu \geq \mu_0$.

We also have to indicate how the perturbed system (14.1), (14.2) converges to the unperturbed one. We assume the following:

$$S_\mu(t)\varphi \to S(t)\varphi, \text{ as } \mu \to \infty, \text{ uniformly with respect to } \varphi \text{ and } t, \text{ in a compact set of } H, t \text{ in a bounded interval.} \tag{14.4}$$

$$\alpha_\mu \geq \alpha_* > 0, \quad \text{for every } \mu \geq \mu_0, \alpha_\mu \to 1 \text{ as } \mu \to \infty. \tag{14.5}$$

$$f_\mu \to f \quad \text{in } \mathcal{D}(A^{1/2}) \text{ as } \mu \to \infty. \tag{14.6}$$

We start with the following technical lemma:

Lemma 14.1. *For every $\varepsilon \geq 0$ there exists $t_\varepsilon < \infty$, t_ε independent of μ, such that Σ_μ is included in an ε-neighborhood of $\Sigma_{\mu\varepsilon}$, where*

$$\Sigma_{\mu\varepsilon} = \bigcup_{0 \leq t \leq t_\varepsilon} S_\mu(t)\Gamma.$$

PROOF. The boundary $\Sigma_{\mu\varepsilon}$ consists of Γ and $S_\mu(t_\varepsilon)\Gamma$. We project on $P_n H$ and use the fact that P_n is obviously a Lipschitz map:

$$\mathrm{vol}_{n-1}(P_n S_\mu(t_\varepsilon)\Gamma) \leq \mathrm{vol}_{n-1}(S_\mu(t_\varepsilon)\Gamma).$$

Then, by Corollary 8.2 for the perturbed system

$$\mathrm{vol}_{n-1}(S_\mu(t_\varepsilon)\Gamma) \leq e^{-kt_\varepsilon}\,\mathrm{vol}_{n-1}(\Gamma) \tag{14.7}$$

where k is some positive constant independent of μ by the uniformity property. Whence the last expression and $\mathrm{vol}_{n-1}(P_n S_\mu(t_\varepsilon)\Gamma)$ can be made arbitrarily small by choosing t_ε sufficiently large. We now denote by ρ, v the radius and center of the largest ball $B_\rho(v)$ of $P_n H$, included in $P_n(\bar{\Sigma}_\mu \backslash \Sigma_{\mu\varepsilon})$:

$$\overline{B_\rho(v)} \subset P_n(\bar{\Sigma}_\mu \backslash \Sigma_{\mu\varepsilon}).$$

Since the boundary of $P_n(\bar{\Sigma}_\mu \backslash \Sigma_{\mu\varepsilon})$ is $P_n(S_\mu(t_\varepsilon)\Gamma)$, the isoperimetric inequality in $P_n H = \mathbb{R}^n$, with appropriate constants $b_n = \omega_n^{-1/n}$, c_n (depending only on n):

$$\rho \leq b_n \operatorname{vol}_n (P_n(\bar{\Sigma}_\mu \backslash \Sigma_{\mu\varepsilon}))^{1/n} \leq c_n \{\operatorname{vol}_{n-1} (S_\mu(t_\varepsilon)\Gamma)\}^{1/(n-1)}.$$

Hence, by (14.7),

$$\rho \leq c_n e^{-kt_\varepsilon/(n-1)} \{\operatorname{vol}_{n-1} (\Gamma)\}^{1/(n-1)}.$$

We will denote by $\eta = \eta_n(t_\varepsilon)$ the right-hand side of the last inequality; obviously $\eta(t_\varepsilon) \to 0$ for $t_\varepsilon \to \infty$. We now want to prove the inclusion

$$\Sigma_\mu \subset B_\varepsilon(\Sigma_{\mu\varepsilon}) = \{u \in H \mid \operatorname{dist}(u, \Sigma_{\mu\varepsilon}) < \varepsilon\}. \tag{14.8}$$

(Note that t_ε is not yet specified.) Let $u = p + \phi_\mu(p)$ be the point of Σ_μ. We can assume that $p + \phi_\mu(p) \notin \Sigma_{\mu\varepsilon}$ since the result is otherwise obvious. Thus $p + \phi_\mu(p) \in \bigcup_{t > t_\varepsilon} S_\mu(t)\Gamma = \bar{\Sigma}_\mu \backslash \Sigma_{\mu\varepsilon}$, $p \in P_n(\bar{\Sigma}_\mu \backslash \Sigma_{\mu\varepsilon})$. By the definition of ρ, and since $\eta \geq \rho$, the ball $B_{2\eta}(p)$ is not included in $P_n(\bar{\Sigma}_\mu \backslash \Sigma_{\mu\varepsilon})$ and therefore there exists $p' \in P_n \Sigma_{\mu\varepsilon}$, with $|p - p'| < 2\eta$. The Lipschitz property of δ_μ then gives

$$|\phi_\mu(p) - \phi_\mu(p')| \leq \tfrac{2}{3}\eta.$$

Thus

$$\operatorname{dist}(u, \Sigma_{\mu\varepsilon}) \leq |u - p' + \phi_\mu(p')| \leq \frac{2\sqrt{10}}{3}\eta.$$

Choosing t_ε such that $\eta(t_\varepsilon) < (3/4\sqrt{10})\varepsilon$, we obtain

$$\operatorname{dist}(u, \Sigma_{\mu\varepsilon}) \leq \frac{\varepsilon}{2} \quad \text{for all } u \in \bar{\Sigma}_\mu \backslash \Sigma_{\mu\varepsilon}.$$

This establishes (14.8) and the lemma too. □

We now state and prove the main result of this section.

Theorem 14.2. *We assume that the conditions in Theorem* 10.1 *are satisfied. Then, as* $\mu \to \infty$,

$$\sup_{p \in D_n} |\phi_\mu(p) - \phi(p)| \to 0.$$

Proof. We assume that $\mu \geq \mu_0$, μ_0 as above, and we are given $\varepsilon > 0$. Lemma 14.1 applies with an appropriate time t_ε. Now if $p \in D_n$, then $p + \phi_\mu(p) \in \Sigma_\mu \subset B_\varepsilon(\Sigma_{\mu\varepsilon})$. This implies the existence of $t_0 = t_0(p, \mu)$, $0 \leq t_0 \leq t_\varepsilon$, and $u_0 \in \Gamma$ such that

$$|p + \phi_\mu(p) - S_\mu(t_0)u_0| \leq \varepsilon.$$

Because of (14.4), there exists μ_ε such that for $\mu \geq \mu_\varepsilon$

$$\sup_{\substack{\varphi \in \Gamma \\ 0 \leq t \leq t_\varepsilon}} |S_\mu(t)\varphi - S(t)\varphi| \leq \varepsilon,$$

whence for $\mu \geq \mu_0$,

$$|S_\mu(t_0)u_0 - S(t_0)u_0| \leq \varepsilon,$$

$$|p + \phi_\mu(p) - S(t_0)u_0| \leq 2\varepsilon.$$

But $S(t_0)u_0$ belongs to Σ and is therefore of the form $p' + \phi(p')$, with some $p' \in D_n$, and we have obtained

$$|p + \phi_\mu(p)| - (p' + \phi(p'))| \leq 2\varepsilon,$$

which implies

$$|\phi_\mu(p) - \phi(p)| \leq |\phi_\mu(p) - \phi(p')| + |\phi(p') - \phi(p)| \leq 2\varepsilon + 2\varepsilon/3 = 8\varepsilon/3.$$

Since p is arbitrary in D_n we conclude that for $\mu \geq \mu_\varepsilon$

$$|\phi_\mu(p) - \phi(p)| \leq 8\varepsilon/3,$$

i.e.,

$$\sup_{p \in D_n} |\phi_\mu(p) - \phi(p)| \leq 8\varepsilon/3.$$

The proof of the theorem is complete. □

Remark 14.3. This convergence theorem applies to the considered perturbations without further proof; as indicated above, the uniformity properties are obtained by a perusal of the discussions and proofs above (the details are left to the reader), while property (14.4) is a classical result in the Galerkin approximation. The situation is slightly complicated by the fact that we introduced the parameter $\alpha_\mu \to 1$, and we allowed for more general perturbations of f than those considered in Galerkin approximations, but this is a totally minor point.

Finally, if one wants to consider more general perturbations and apply Theorem 14.2 to them, the necessary technical material is (14.4) to (14.6) and the uniformity of the Lipschitz constant $l = \frac{1}{3}$ of ϕ_μ and that of the domain D_n on which δ_μ are defined (for μ large enough); the proof of Theorem 14.2 is otherwise unchanged.

CHAPTER 15

Application: The Kuramoto–Sivashinsky Equation

We recall that in the case of the Kuramoto–Sivashinsky [HN, HN1, HNZ, NSh] equation on the space H of odd L-periodic functions, $(du/dt) + Au + R(u) = 0$, we have

$$R(u) = B(u, u) + Cu + f, \tag{15.1}$$

with

$$B(u, v) = u\frac{dv}{dx},$$

$$Cu = -A^{1/2}u + B(u, \varphi) + B(\varphi, u),$$

$$f = A\varphi + \psi \quad \text{with } \psi = \frac{d^2\varphi}{dx^2} + \varphi\frac{d\varphi}{dx},$$

with the explicit time-independent φ defined in [FNST, FNST1],

$$\Lambda_n = \lambda_n = c_0\left(\frac{n}{L}\right)^4, \qquad n = 1, 2, \dots.$$

(Here as in the sequel c_0, c_1, ... denote absolute constants; for instance, $c_0 = (2\pi)^4$.) Also we shall consider $L \geq 1$, the case $L < 1$ being of no interest.

The coercitivity condition (4.5) is true with a constant $k_4 = c_1$ *independent of* L (see [NST, NST1]). Also there exists a constant c_2 independent of L such that $|A^{1/2}\varphi|^2 + |\psi|^2 \leq c_2^2 L^5$ [NST1]. Therefore the condition on R, the radius of the set of initial data Γ (see Proposition 9.2), becomes

$$R = \beta_1 L^{5/2}, \qquad \beta_1 \geq \frac{4c_2}{c_1}. \tag{15.2}$$

Hereafter β_1 as well as β_2, β_3, ..., appearing below are parameters that will be chosen in a consistent way at the end of our discussion.

So $\Gamma = \{u \mid P_n u = u, |u| = R\}$, R given in (15.2). We now proceed to determine the condition imposed on R and n by the other requirement (9.11) in Proposition 9.2. For the convenience of the reader we reproduce (9.1) here:

$$\frac{|u|^2 |A^{1/2}(I - P_n)N(u)|^2}{(|A^{1/2}u|^2 + |u|^2)^2} \leq \frac{c_1^2}{8}(\lambda_{n+1} - \lambda_n) \quad \text{for } u \in \Gamma. \tag{15.3}$$

We observe that

$$(I - P_n)(N(u)) = (I - P_n)[B(u,u) + B(u,\varphi) + B(\varphi,u) + f].$$

Now, it is known [NST1] that φ lies in a space $P_M H$ with M of the order of L^2, $M = [C_3 L^2]$. Therefore f will lie in the space $P_{2M}H$. We can thus ignore the contribution from f:

$$(I - P_n)f = 0 \quad \text{if } n \geq c_3 L^2.$$

So $(I - P_n)N(u) = (I - P_n)[B(u,u) + B(u,\varphi) + B(\varphi,u)]$. From expression (15.1) and $P_n u = u$, $P_n = \varphi$ it follows that $[B(u,u) + B(u,\varphi) + B(\varphi,u)] \in P_{2n}H$. Therefore

$$\begin{aligned} |A^{1/2}(I - P_n)N(u)| &\leq (\lambda_{2n})^{1/2} |((I - P_n)[B(u,u) + B(u,\varphi) + B(\varphi,u)]| \\ &\leq \lambda_{2n}^{1/2}\lambda_{n+1}^{-1/4} |A^{1/4}[B(u,u) + B(u,\varphi) + B(\varphi,u)]|. \end{aligned}$$

Now

$$\begin{aligned} &|A^{1/4}(B(u,u) + B(u,\varphi) + B(\varphi,u))|^2 \\ &\leq 3\left(\left|\frac{d}{dx}\left(u\frac{du}{dx}\right)\right|^2 + \left|\frac{d\varphi}{dx}\left(u\frac{d\varphi}{dx}\right)\right|^2 + \left|\frac{d}{dx}\left(\varphi\frac{du}{dx}\right)\right|^2\right). \end{aligned}$$

We estimate

$$\left|\frac{d}{dx}\left(u\frac{du}{dx}\right)\right|^2 \leq 2\left(\left|u\frac{d^2u}{dx^2}\right|^2 + \left|\left(\frac{du}{dx}\right)^2\right|^2\right),$$

$$\int_0^L u^2\left(\frac{d^2u}{dx^2}\right)^2 dx \leq |u|_{L^\infty}^2 |A^{1/2}u|^2 \leq |u|^{3/2}|A^{1/2}u|^{5/2}.$$

Also the term

$$\begin{aligned} \int_0^L \left(\frac{du}{dx}\right)^4 dx &= -3\int_0^L u\left(\frac{du}{dx}\right)^2 \frac{d^2u}{dx^2}dx \\ &\leq \frac{1}{2}\int_0^L \left(\frac{du}{dx}\right)^4 dx + \frac{9}{2}\int_0^L u^2\left(\frac{d^2u}{dx^2}\right)^2 dx, \end{aligned}$$

thus

$$\int_0^L \left(\frac{du}{dx}\right)^4 \leq 9|u|^{3/2}|A^{1/2}u|^{5/2}.$$

We obtained

$$\left|\frac{d}{dx}\left(u\frac{du}{dx}\right)\right|^2 \le 20|u|^{3/2}|A^{1/2}u|^{5/2}.$$

We now have to estimate the integrals

$$I_1 = \int_0^L \left(\frac{du}{dx}\right)^2\left(\frac{d\varphi}{dx}\right)^2 dx, \quad I_2 = \int_0^L u^2\left(\frac{d^2\varphi}{dx^2}\right)^2 dx, \quad I_3 = \int_0^L \varphi^2\left(\frac{d^2u}{dx^2}\right)^2 dx.$$

We shall make use of the fact (see [NST1]) that $|\varphi|_{L^\infty} \le c_4 L$ and $|A^{1/2}\varphi|^2 = \int_0^L (d^2\varphi/dx^2)^2\, dx \le c_1^2 L^5$:

$$\begin{aligned}
I_1 + \int_0^L \left(\frac{du}{dx}\right)^2\left(\frac{d\varphi}{dx}\right)^2 &\le \left(\int_0^L \left(\frac{du}{dx}\right)^4\right)^{1/2}\left(\int_0^L \left(\frac{d\varphi}{dx}\right)^4\right)^{1/2} \\
&\le 9|u|^{3/4}|A^{1/2}u|^{5/4}\left(\int_0^L \varphi^2\left(\frac{d^2\varphi}{dx^2}\right)^2 dx\right)^{1/2} \\
&\le c_5 L^{7/2}|u|^{3/4}|A^{1/2}u|^{5/4},
\end{aligned}$$

and similarly,

$$I_2 = \int_0^L u^2\left(\frac{d^2\varphi}{dx^2}\right)^2 dx \le c_2^2|u|^{3/2}|A^{1/2}u|^{1/2}L^5,$$

$$I_3 = \int_0^L \varphi^2\left(\frac{d^2u}{dx^2}\right)^2 dx \le c_4^2 L^2|A^{1/2}u|^2.$$

Summing up, (9.11) will be satisfied if

$$\frac{n^2}{L^2}|u|^2\frac{(|u|^{3/2}|A^{1/2}u|^{5/2} + L^2|A^{1/2}u|^2 + L^5|u|^{3/2}|A^{1/2}u|^{1/2} + L^{7/2}|u|^{3/4}|A^{1/2}u|^{5/4})}{(|A^{1/2}u|^2 + |u|^2)^2}$$

$$\le c_6\frac{n^3}{L^4},$$

where we used $\lambda_{2n}\lambda_{n+1}^{-1/2} \le 16c_0^{1/2}(n^2/L^2)$, $\lambda_{n+1} - \lambda_n \ge 4c_0(n^3/L^4)$. We write the last inequality in the form

$$\begin{aligned}
n \ge \frac{1}{c_6}\Bigg(& L^2|u|^2\frac{|u|^{3/2}|A^{1/2}u|^{5/2}}{(|A^{1/2}u|^2 + |u|^2)^2} + L^4\frac{|u|^2|A^{1/2}u|^2}{(|A^{1/2}u|^2 + |u|^2)^2} \\
&+ L^7\frac{|u|^{7/2}|A^{1/2}u|^{1/2}}{(|A^{1/2}u|^2 + |u|^2)^2} + L^{11/2}\frac{|u|^{11/4}|A^{1/2}u|^{5/4}}{(|A^{1/2}u|^2 + |u|^2)^2}\Bigg).
\end{aligned}$$

Now by Young's inequality, all the ratios on the right-hand side of this relation are bounded by absolute constants. Thus (15.3) will be satisfied if

$$n \ge c_7(L^2|u|^2 + L^4 + L^7 + L^{11/2}).$$

Since, if $u \in \Gamma$, $|u| = R = \beta_1 L^{5/2}$, we obtain that (15.3) will be satisfied if $n \ge \max\{c_3 L^2, c_7(\beta_1^2 + 3)L^7\}$. Since $L \ge 1$ we summarize the results obtained up to now as follows:

Lemma 15.1. *In the case of equation* (15.1), *if*

$$\Gamma = \{u \mid P_n u = u, |u| = R\},$$

with $R = \beta_1 L^{5/2}$ *and* $n \geq \beta_2 L^7$ *where*

$$\beta_1 \geq \frac{4c_2}{c_1} \quad \text{and} \quad \beta_2 \geq c_3 + c_7(\beta_1^2 + 3), \tag{15.4}$$

then condition (C1) *in Chapter* 10 *is satisfied, provided*

$$R \geq \rho_0 + \frac{4\rho_1}{\lambda_{n+1}} \quad (cf.\ (6.3)). \tag{15.5}$$

In what follows, we shall show that for β_2 large enough the condition

$$n \geq \beta_2 L^7 \tag{15.6}$$

is sufficient to ensure (15.5) the decay of volume elements, spectral blocking, and the strong squeezing properties required in the geometric arguments of Chapter 10, i.e., conditions (C2) to (C5) in Chapter 10.

We recall first that by the choice of $R = \beta_1 L^{5/2}$ we have that the ball

$$D = \{u \in H : |u| \leq R\} \tag{15.7}$$

is invariant for (15.1) (see (9.3)). We shall now define

$$Y = D \in \{u \in \mathscr{D}(A^{1/4}) \mid |A^{1/2}u| \leq \beta_3 L^{8.5}\} \tag{15.8}$$

with an absolute constant β_3 to be estimated later, after Lemma 15.2. This choice will make $\theta = 1$ in condition (C2) (see Chapter 10). In order to obtain the necessary constraints on β_1 and β_3 we consider a $u_0 \in D$ such that $|A^{1/4}u_0| \leq \beta_3 L^{8.5}$.

We recall that the change of variables $u + \varphi = v$ brings the equation to its original form:

$$\begin{cases} \dfrac{\partial}{\partial t} v + \dfrac{\partial^4}{\partial x^4} v + \dfrac{\partial^2}{\partial x^2} v + v\dfrac{\partial v}{\partial x} = 0, \\ v(0) = v_0 \end{cases}$$

with v an odd periodic function of period L. From the estimates on φ it follows that $|v(t)| \leq (\beta_1 + c_4)L^{5/2}$ for $t \geq 0$. The equation for $|dv/dx|$ is obtained by multiplying the Kuramoto–Sivashinsky equation by $-d^2v/dx^2$ and integrating by parts:

$$\frac{1}{2}\frac{d}{dt}\left|\frac{dv}{dx}\right|^2 = \left|\frac{d^3x}{dx^3}\right|^2 - \left|\frac{d^2v}{dx^2}\right|^2 - \int_{-L/2}^{L/2} v(x)\frac{dv}{dx}\frac{d^2v}{dx^2}\,dx = 0.$$

We estimate the cubic term as follows:

$$\left|\int_{-L/2}^{L/2} v(x)\frac{dv}{dx}\frac{d^2v}{dx^2}\,dx\right| \leq |v|_{L^\infty}\left|\frac{dv}{dx}\right|\left|\frac{d^2v}{dx^2}\right| \leq |v|^{11/6}\left|\frac{d^3v}{dx^3}\right|^{7/6}.$$

On the other hand,

$$\left|\frac{d^3v}{dx^3}\right|^2 - \left|\frac{d^2v}{dx^2}\right|^2 \geq \left|\frac{d^2v}{dx^2}\right|^2 - \left|\frac{dv}{dx}\right|^2 \geq \left|\frac{dv}{dx}\right|^2 - |v|^2$$

and

$$\left|\frac{d^2v}{dx^2}\right|^2 \leq |v|^{2/3}\left|\frac{d^3v}{dx^3}\right|^{4/3} \leq \frac{1}{2}\left|\frac{d^3v}{dx^3}\right|^2 + \frac{16}{27}|v|^2,$$

thus

$$\begin{aligned}\left|\frac{d^3v}{dx^3}\right|^2 - \left|\frac{d^2v}{dx^2}\right| &\geq \frac{1}{2}\left(\left|\frac{d^3v}{dx^3}\right|^2 - \left|\frac{d^2v}{dx^2}\right|^2\right) + \frac{1}{2}\left|\frac{dv}{dx}\right|^2 - \frac{1}{2}|v|^2 \\ &\geq \frac{1}{4}\left|\frac{d^3v}{dx^3}\right|^2 + \frac{1}{2}\left|\frac{dv}{dx}\right|^2 - \frac{3}{4}|v|^2.\end{aligned}$$

The cubic term can be estimated further:

$$\left|\int_{-L/2}^{L/2} v\frac{dv}{dx}\frac{d^2v}{dx^2}dx\right| \leq 2|v|^{11/6}\left|\frac{d^3v}{dx^3}\right|^{7/6} \leq \frac{1}{2}\left|\frac{d^3v}{dx^3}\right|^2 + 2^{7/5}|v|^{22/5}.$$

We obtained

$$\frac{1}{2}\frac{d}{dt}\left|\frac{dv}{dx}\right|^2 + \frac{1}{2}\left|\frac{dv}{dx}\right|^2 - |v|^2 - 3|v|^{22/5} \leq 0. \tag{15.9}$$

Since we have the bound

$$\tfrac{3}{2}|v|^2 + \tfrac{3}{2}|v|^{22/5} \leq \beta_4 L^{11}, \qquad \beta_4 = (\beta_1 + c_4)^2 + 3(\beta_1 + c_4)^{22/5},$$

we infer that

$$\frac{d}{dt}\left|\frac{dv}{dx}\right|^2 + \left|\frac{dv}{dx}\right|^2 - 2\beta_4 L^{11} \leq 0.$$

Therefore if $|(dv/dx)| \leq \beta_5 L^{8.5}$ at $t = 0$, with $\beta_5 \geq (2\beta_4)^{1/2}$, it will stay smaller than $\beta_5 L^{8.5}$ for all $t \geq 0$. Since the estimate on $|(d\varphi/dx)| = |A^{1/4}\varphi|$ is much better than $L^{8.5}$ (namely $|A^{1/4}\varphi| \leq c_2^{1/2}c_4^{1/2}L^2$), we deduce the first statement in the following:

Lemma 15.2. *If u_0 satisfies $|u_0| \leq \beta_1 L^{5/2}$, $|A^{1/4}u_0| \leq \beta_3 L^{8.5}$, where β_1 satisfies* (15.4) *and*

$$\beta_3 + c_8 \geq (2\beta_4)^{1/2}, \qquad \beta_4 = (\beta_1 + c_4)^2 + 3(\beta_1 + c_4)^{22/5}, \tag{15.10}$$

with $c_8 = c_2^{1/2}c_4^{1/2}$, then the solution $u(t)$ of (15.1), *with u_0 as initial data, satisfies $|A^{1/4}u(t)| \leq (\beta_3 + c_8)L^{8.5}$ for all $t \geq 0$. Moreover,*

$$\limsup_{t\to\infty}\left|\frac{du}{dx}\right| \leq c_7 L^{5.5}. \tag{15.11}$$

PROOF. It remains to prove (15.11). We first notice that by virtue of (9.3) we have

$$\limsup_{t\to\infty} |u(t)| \le \frac{2c_2}{c_1} L^{5/2}, \tag{15.12}$$

hence

$$\limsup |v(t)| \le \left(\frac{2c_2}{c_1} L^{5/2} + c_6 L^{3/2}\right) \le \left(\frac{2c_2}{c_1} + c_6\right) L^{5/2}.$$

Relation (15.11) follows by integrating (15.9), taking into account the above relation and setting

$$c_7 = c_6 + \left[\frac{3}{4}\left(\frac{2c_2}{c_1} + c_6\right)^2 + \frac{3}{2}\left(\frac{2c_2}{c_1} + c_6\right)^{22/5}\right]^{1/2}. \qquad \square$$

Now if β_3 satisfies (15.10) then it is clear that, by virtue of Lemma 15.2, Y satisfies (5.1) and (5.3). Also, since

$$|A^{1/4}u| \le \lambda_n^{1/4}|u| \le c_0^{1/4}\frac{n}{L}\beta_1 L^{5/2} = c_0^{1/4}\beta_1\beta_2 L^{8.5} \quad \text{for } u \in \Gamma,$$

we infer that condition (C2), with $\theta = 1$, in Chapter 10 holds for our present Γ and Y provided

$$c_0^{1/4}\beta_1\beta_2 \le \beta_3. \tag{15.13}$$

Also from (15.11), (15.12) it easily follows that for the global attractor X of (15.1) we have

$$\rho_1 = \sup_{u\in X} |A^{1/4}u| \le c_7 L^{5.5}, \qquad \rho_0 = \sup_{u\in X} |u| \le \frac{2c_2}{c_1} L^{2.5}.$$

Therefore, by Lemma 15.2, condition (C1) is satisfied provided (cf. (15.5))

$$\beta_1 \ge \frac{2c_2}{c_1} + \frac{4c_2}{c_0^{1/4}\beta_2 L^{1/2}} \ge \frac{2c_2}{c_1} + \frac{4c_2 L^{6.5}}{c_0^{1/4} n};$$

in particular, since β_1 satisfies (15.4) and $L \ge 1$, if

$$\beta_2 \ge \frac{2c_1}{c_0^{1/4}}, \tag{15.13a}$$

then condition (C1) in Chapter 10 is satisfied.

Our next step is to verify that our Y has the property (5.2) and to check that the assumptions $N \sim L^7$ and $R \sim L^{5/2}$ are sufficient to ensure conditions (C2) and (C3) in Chapter 10. We remark that the linearized operators $L(t)$ and their adjoints have the form

$$L(t)w = -A^{1/2}w + B(u(t), w) + B(w, u(t)) + B(\varphi, w) + B(w, \varphi), \tag{15.14}$$

$$L(t)^*w = -A^{1/2}w - B(u(t), w) - B(\varphi, w). \tag{15.15}$$

We shall give explicit estimates on k_1, k_2, k_3 using the form (15.14), (15.15) of the operators and the information in φ and in Lemma 15.2. In

$$|L(t)w|^2 \leq 5[|A^{1/2}w|^2 + |B(u(t), w)|^2 + |B(w, u(t))|^2 + |B(\varphi, w)|^2 + |B(w, \varphi)|^2],$$

$$|L(t)^*w|^2 \leq 3[|A^{1/2}w|^2 + |B(u(t), w)|^2 + |B(\varphi, w)|^2],$$

we estimate the terms as follows:

$$\begin{aligned} |B(w, u(t))|^2 &\leq |w|^2_{L^\infty}|A^{1/4}u(t)|^2 \leq (\beta_3 + c_8)^2 L^{17}|w||A^{1/4}w| \\ &\leq \tfrac{1}{2}(\beta_3 + c_8)^2 L^{17}|w|^2 + \tfrac{1}{2}(\beta_3 + c_8)^2 L^{17}|A^{1/4}w|^2; \\ |B(u(t), w)|^2 &\leq |u(t)|^2_{L^\infty}|A^{1/4}w|^2 \leq |u(t)||A^{1/4}u(t)||A^{1/4}w|^2 \\ &\leq \beta_1(\beta_3 + c_8)L^{11}|A^{1/4}w|^2; \\ |B(\varphi, w)|^2 &\leq |\varphi|^2_{L^\infty}|A^{1/4}w|^2 \leq c_4^2 L^2|A^{1/4}w|^2; \\ |B(w, \varphi)|^2 &\leq |A^{1/4}\varphi|^2|w||A^{1/4}w| \leq \frac{c_2^2}{2}L^5|w|^2 + \frac{c_2^2}{2}L^5|A^{1/4}w|^2. \end{aligned}$$

Collecting the estimates we obtain (3.7), (3.8) with

$$\begin{cases} k_1 = 5(\beta_3 + c_8)^2 L^{17} + \tfrac{5}{2}c_2^2 L^5, \\ k_2 = 2k_1 + 5\beta_1^2 L^5 + 5c_4^2 L^2, \\ k_3 = 5. \end{cases} \tag{15.16}$$

We have to estimate the constants k_5 and k_6 in the estimate (4.5a), i.e.,

$$|(B(u, w) + B(w, u), w)| \leq k_5|w|^{3/2}|A^{1/2}w|^{1/2} + k_6|w|^2,$$

where u satisfies $|u| \leq \beta_1 L^{5/2}$, $|A^{1/4}u| \leq (\beta_3 + c_8)L^{8.5}$. But $(B(u, w) + B(w, u), w) = -(B(u, w), w)$, so (4.5a) holds provided

$$k_5 = (2\beta_1(\beta_3 + c_8))^{1/2}L^{8.5}, \qquad k_6 = 0. \tag{15.17}$$

Also, we obtain in the same way that relation (5.5), i.e.,

$$|(L(t)w, w)| = ||A^{1/4}w|^2 + (B(u, w), w) + (B(\varphi, w), w)| \leq k_9|w|^2 + k_9'|A^{1/4}w|^2,$$

is satisfied with

$$k_9 = \tfrac{1}{2}\beta_1(\beta_3 + c_8)L^{11} + \tfrac{1}{2}c_4^2 L^2, \qquad k_9' = \tfrac{5}{2}. \tag{15.18}$$

By replacing β_3 in (15.16), (15.17), and (15.18) with $\theta\beta_3$ we immediately see that Y has property (5.2) as well as the property completing (5.5).

We are now in a position to check conditions (C4) and (C3) in Chapter 10. These conditions are, in order,

$$\left(\frac{\lambda_{n+1} - \lambda_n}{2}\right)^2 > k_1 + k_2\left(\frac{\lambda_{n+1} + \lambda_n}{2}\right)^{1/2} + k_3\frac{\lambda_{n+1} + \lambda_n}{2}, \tag{15.18a}$$

$$\lambda_{n+1} + \lambda_n \geq 3k_5^{4/3}/(3c_1)^{1/3}, \tag{15.18b}$$

$$\lambda_{n+1} - \lambda_n \geq k_9\lambda_n^{1/2} + \tfrac{13}{3}(k_1' + k_2\lambda_n^{1/2} + k_3\lambda_n)^{1/2}, \tag{15.18c}$$

where $k_1' = k_1 + k_9'^2$.

Relation (15.18b) is ensured if

$$\frac{c_0 n^4}{L^4} \geq \frac{3}{(2c_1)^{1/3}} [2\beta_1(\beta_3 + c_8)]^{2/3} L^{22/3}$$

or even more if

$$\beta_2^4 \geq \frac{3}{c_0(2c_1)^{1/3}} [2\beta_1(\beta_3 + c_8)]^{2/3}. \tag{15.18d}$$

Both relations (15.18a), (15.18c) obviously follow from

$$\left(\frac{\lambda_{n+1} - \lambda_n}{2}\right)^2 \geq c_9 k_1' + c_9 k_2 \lambda_{n+1}^{1/2} + (c_9 k_3 + 2k_9^2)\lambda_{n+1},$$

where $c_9 = \frac{338}{9}$. By using (15.16), (15.17), $L \geq 1$ and by introducing new adequate constants c_{10} to c_{13} the above relation can be given the following strengthened form:

$$\frac{4c_0^2 n^6}{L^8} \geq c_{10}[\beta_1^2(\beta_3 + c_8)^2 + c_4^4]L^{22}$$
$$+ c_{11}[(\beta_3 + c_8)^2 + \beta_1^2 + c_{12}]L^{17} \frac{c_0^{1/2}(n+1)^2}{L^2} + c_{13} \frac{c_0(n+1)^4}{L^4},$$

which is satisfied if, for instance,

$$n^6 \geq (3c_{10}/4c_0^2)[\beta_1^2(\beta_3 + c_8)^2 + c_4^4]L^{30},$$
$$n^4 \geq 4(3c_{11}/4c_0^{3/2})[(\beta_3 + c_8)^2 + \beta_1^2 + c_{12}]L^{15},$$
$$n^2 \geq 16(3c_{13}/4c_0)L^4.$$

Since $L \geq 1$, these relations will be satisfied by $n = \beta_2 L^7$ if, with an obvious choice of new constants c_{14}, c_{15}, we have that

$$\beta_2^3 \geq c_{14}[(\beta_3 + c_8)^2 + \beta_1^2 + c_{15}]. \tag{15.19}$$

Finally, let us check the last condition (C5) in Chapter 10, namely

$$\lambda_1 + \cdots + \lambda_m > (2k_9' + k_9^2)m, \quad \text{for } m \geq n - 2.$$

For this, since

$$\lambda_1 + \cdots + \lambda_m \geq \frac{c_0}{L^4}\frac{m^5}{5},$$

it is sufficient that

$$\frac{c_0}{L^4}\frac{(n-2)^4}{5} > 2k_9' + k_9^2,$$

that is, with a new appropriate constant c_{16}

$$\beta_2 \geq \left(\frac{5}{c_0}\right)^{1/4} [\beta_1(\beta_3 + c_8) + c_{16}]^{1/4}. \tag{15.20}$$

Increasing in an obvious way the constants c_{14} and c_{15} in (15.19), we can consider that (15.19) implies (15.18d) and (15.20).

We have thus proved that if the parameters β_1, β_2, and β_3 satisfy relations (15.4), (15.10), (15.13), (15.13a), and (15.19) then all conditions (C1) to (C5) in Chapter 10 are satisfied too. We now choose $\beta_1 = c_{17} = 4c_2/c_1$. Then the second relation (15.4) and (15.13a), (15.10), and (15.13) become, in order,

$$\beta_2 \geq c_{18}, \qquad \beta_3 \geq c_{19}, \qquad \beta_3 \geq c_{10}\beta_2, \tag{15.21}$$

while relation (15.19) becomes

$$\beta_2 \geq c_{21}\beta_3^{2/3} + c_{22}. \tag{15.22}$$

Taking $\beta_3 = c_{20}\beta_2$, relation (15.22) takes the form

$$\beta_2 \geq c_{23}\beta_2^{2/3} + c_{22},$$

which is satisfied provided $\beta_2 \geq c_{24}$.

In conclusion, Theorem 10.1 can be applied and we obtain Theorem 15.3.

Theorem 15.3. *Let Γ be the $(n-1)$-dimensional sphere in P_nH with center* 0 *and of radius $R = c_{16}L^{5/2}$ and $n \geq c_{25}L^7$ (where c_{16} and $c_{25} \geq \max\{c_{24}, c_{18}, c_{19}/c_{20}\}$ are appropriate absolute constants). Then the closure $\bar{\Sigma}$ of the integral manifold for equation* (15.1) *constructed with Γ as initial data is an inertial manifold for* (15.1), *having all properties listed in Theorem* 10.1.

The natural question to ask now is under what supplementary conditions does Theorem 12.1 hold for the inertial manifold $\bar{\Sigma}$ considered in Theorem 15.3. It is an easy exercise to check that the second supplementary condition in Theorem 15.3 is satisfied if the constant c_{25} in Theorem 15.3 is large enough. (Indeed, one can prove that $k_{14} = \frac{1}{2} + \frac{3}{4}c_4^{4/3}L^{4/3}$, $k_{15} = 1$ will do in (13.14).) The first supplementary condition in Theorem 12.1, namely

$$\lambda_n + \lambda_{n+1} \geq k_3'^2/2 = (k_1^{1/2} + \tfrac{1}{4}k_2^{1/2})^2/2,$$

will be satisfied if

$$\frac{2c_0 n^4}{L^2} \geq k_1 + \tfrac{1}{4}k_2,$$

that is (by (15.16)),

$$2c_0\beta_2^4 L^{24} \geq c_{26}[(\beta_2 + c_8)^2 + c_{27}]L^{17}.$$

Finally, since $L \geq 1$ this relation is obviously implied by (15.19) provided c_{14} and c_{15} are chosen large enough. Thus from Theorem 12.1 we deduce Theorem 15.4.

Theorem 15.4. *If in Theorem* 15.3, c_{25} *is taken large enough, then the inertial manifold* $\bar{\Sigma}$ *is asymptotically complete.*

We conclude this chapter by mentioning that in [FNST1] one constructs inertial manifolds of dimension $n \sim L^{3.5}$ for L large enough. However, the construction in this work yields a much more precise description of the inertial manifolds $\bar{\Sigma}$ as well as of their direct connection to equation (15.1).

CHAPTER 16

Application: A Nonlocal Burgers Equation

The equation we will investigate here is of the form

$$\frac{du}{dt} + N(u) = 0, \qquad N(u) = Au + R(u), \tag{16.1}$$

on $H = \{u \in L^2(0, L) : \int_0^L u(x)\,dx = 0, 0 \le x \le L\}$, where

$$\begin{cases} R(u) = B(u, u) + f, \qquad \varphi = 0, \psi = f \in H, f \neq 0, \\ B(u, v) = (u, \omega)v', \quad \text{where } v' = dv/dx, \text{ with a fixed } \omega \in H, \omega \neq 0, \\ A = -\dfrac{d^2}{dx^2} \quad \text{with periodic boundary conditions.} \end{cases} \tag{16.1a}$$

Precisely, the domain of definition of A will be $\mathcal{D}(A) = \{u \in H^2(0, L) \mid u(0) = u(L), u'(0) = u'(L)\}$, where $H^2(0, L)$ denotes the usual L^2-Sobolev space of order 2; obviously $\Lambda_n = \lambda_{2n} = \lambda_{2n-1} = (2\pi n/L) = \lambda_1 n^2, n = 1, 2, \ldots$. We shall also assume that $f \in \mathcal{D}(A)$.

The general existence and uniqueness theory for this equation is trivial. We note only that any solution $u = u(t)$ with initial data u_0 satisfies

$$\begin{cases} |u|^2 \le |u_0|^2 e^{-\lambda_1 t} + \dfrac{|f|^2}{\lambda_1^2}(1 - e^{-\lambda_1 t}), \\ |A^{1/2}u|^2 \le |A^{1/2}u_0|^2 e^{-\lambda_1 t} + \dfrac{|f|^2}{\lambda_1}(1 - e^{-\lambda_1 t}), \end{cases} \tag{16.2}$$

which follows easily from

$$\begin{cases} (B(u, v), v) = 0 \quad \text{for } u \in H, v \in \mathcal{D}(A^{1/2}), \\ (B(u, w), Aw) = 0 \quad \text{for } u \in H, w \in \mathcal{D}(A). \end{cases} \tag{16.3}$$

These identities are due to the fact that $v(0)^2 = v(L)^2$, $w'(0)^2 = w'(L)^2$ for $f \in \mathscr{D}(A^{1/2})$, $w \in \mathscr{D}(A)$.

We now consider $R_0 \geq |f|/\lambda_1$, $R_1 \geq |f|/\lambda_1^{1/2}$ and consider *only* solutions $u = u(t)$ with initial data u_0 satisfying $|u_0| \leq R_0$, $|u_0'| = |A^{1/2}u_0| \leq R_1$. By virtue of (16.2) we have

$$|u(t)| \leq R_0, \qquad |A^{1/2}u(t)| \leq R_1, t \geq 0. \tag{16.4}$$

The operators $L(t)$ and $L(t)^*$ are now given by

$$\begin{cases} L(t)g = B(u,g) + B(g,u), \\ L(t)^*g = -B(u,g) + (g,u')\omega. \end{cases} \tag{16.5}$$

So the relations (3.7), (3.8) along the solutions $u = u(t)$ satisfying (16.4) are verified if k_1, k_2, k_3 are chosen according to the following formula:

$$k_1 = 2|\omega|^2 R_1^2, \qquad k_2 = 0, \qquad k_3 = 2|\omega|^2 R_0^2. \tag{16.6}$$

But in this case condition (3.13) in Chapter 3 can be replaced by

$$\left(\frac{\Lambda_{m+1} - \Lambda_m}{2}\right)^2 > k_1 = 2|\omega|^2 R_1^2. \tag{16.7}$$

Indeed, in this case in the basic equations (3.4), (3.6) the terms $((A - \Lambda)g, (I - P)Lg)$, resp. $((A - \lambda)h, PL^*h)$ can, by virtue of (16.3), be replaced by

$$((A - \Lambda)g, (I - P)B(g,u)), \quad \text{resp. } ((A - \lambda)h, (h,u')P\omega).$$

Therefore by virtue of Chapter 3, the spectral blocking properties, along the solutions satisfying (16.4), are valid provided (16.7) holds.

In order to check the strong squeezing properties (Theorems 4.2 and 4.3) for solutions satisfying (16.4), we first determine the constants k_5, k_6 along any solution u satisfying (16.4). Indeed, since

$$|(B(u,w) + B(w,u), w)| = |(B(w,u),w)| \leq |\omega||w|^2|u'| \leq R_1|\omega||w|^2$$

we can take

$$k_5 = 0, \qquad k_6 = R_1|\omega|, \qquad k_7 = 2k_6. \tag{16.8}$$

Hence the constant k_7 in (4.7) is $2k_6$. Second, in the present case the coercivity condition (4.5) can obviously be replaced with the condition

$$(Ag,g) \geq \lambda_1|g|^2 \quad \text{for all } g \in \mathscr{D}_A. \tag{16.9}$$

Also, in the equation for $|w|^2$ (see Chapter 4 between (4.5a) and (4.6)) the k_4 can be taken $= 1$. Therefore condition (4.6) becomes

$$\Lambda_m + \Lambda_{m+1} > 2k_7 = 4|\omega|R_1. \tag{16.10}$$

Thus, by virtue of Chapter 4, the strong squeezing properties provided by Theorems 4.2 ad 4.3 for the solutions satisfying (16.4) are valid provided (16.7) and (16.10) hold.

For the set Y in Chapter 5 we take

$$
\begin{aligned}
Y &= \left\{ u \in \mathscr{D}(A^{1/2}) : |u| \le \frac{2|f|}{\lambda_1}, |A^{1/2}u| \le \frac{2|f|}{\lambda_1^{1/2}} \right\} \\
&= \left\{ u \in \mathscr{D}(A^{1/2}) : |A^{1/2}u| \le \frac{2|f|}{\lambda_1^{1/2}} \right\}.
\end{aligned}
\tag{16.11}
$$

Then obviously Y has the property (5.1) (with the power $A^{1/4}$ replaced by the power $A^{1/2}$) and the properties (5.2) (by our previous computation of the constants k_1, k_2, k_3, k_5, k_6) and (5.3) (by (16.2)). Condition (5.5), and the requirement of its validity for all solutions starting from some θY (with $\theta \ge 1$), follows from

$$
(L(t)w, w) = |(B(w, u(t)), w)| \le k_9 |w|^2
$$

with $k_9 = R_1 |\omega|$ for all solutions $u(t)$ satisfying (16.4) (i.e., $u_0 \in \theta Y$, with $\theta = \max\{R_0 \lambda_1/|f|, R_1 \lambda_1^{1/2}/|f|\}$). In the present case, perusal of the proof of Theorem 5.1 shows that condition (5.6) can be replaced by

$$
\lambda_{n+1} - \lambda_n > \frac{1 + \gamma^2}{\gamma} |\omega| R_1. \tag{16.12}
$$

For $\gamma = \frac{1}{3}$ this becomes

$$
\lambda_{n+1} - \lambda_n > \tfrac{10}{3} |\omega| R_1. \tag{16.12a}
$$

Applying these facts to the global attractor X (for which the values $R_0 = |f|/\lambda_1$, $R_1 = |f|/\lambda_1^{1/2}$ are adequate), we deduce as in Chapter 6 (see Remark 6.1a) the following:

Proposition 16.1. *For*

$$
m > 2^{1/2} r \quad \text{and} \quad r = |\omega||f|/\lambda_1^{3/2} \ge 1, \tag{16.13}
$$

$P_{2m} | X$ *is injective and has a Lipschitz inverse.*

We notice only that (16.13) and $r \ge 1$ imply (16.7) and (16.10) and that $r < 1$ implies that $X = \{u_0\}$ where u_0 is in this case the unique stationary solution. Therefore we shall consider in the sequel that $r \ge 1$. Moreover, for any orthogonal projection P of rank l and any solution $u(t)$ satisfying (16.4) we have (see Chapter 8)

$$
\operatorname{Tr}(A(t)P) \ge 2\lambda_1 \sum_1^j k^2 - |\omega||u'(t)| \ge \lambda_1 \left(\frac{j(j + 1/2)(j - 1)}{3} - \frac{|\omega| R_1}{\lambda_1} \right),
$$

where $j = l/2$ if $l =$ even and $j = (l + 1)/2$ if $l =$ odd. Taking $R_1 = |f|/\lambda_1^{1/2}$ here, an application of the proof of Theorem 8.4 (by using $\overline{q}_j = \lambda_1(\frac{1}{3} j(j + \frac{1}{2}) \times (j + 1) - r)$ gives us the following supplement to Proposition 16.1:

Proposition 16.2. *The Hausdorff dimension and the fractal dimension, $d_H(X)$ and $d_M(X)$, of X satisfy*

$$d_H(X) \leq 2m_0, \qquad d_M(X) \leq 4m_1, \tag{16.14}$$

where m_0 and m_1 are the first integers such that $m^3 > 3r$ and $m^3 > 6r$, respectively.

Finally, in connection with Chapter 8 we mention that the validity condition in Corollary 8.2 is assumed by the inequality

$$m > 2(3r)^{1/3}. \tag{16.15}$$

We will now outline the main adaptation of the theory of inertial manifolds developed in Chapters 9 to 13. We will take

$$\Gamma = \{u \in P_nH : |A^{1/2}u|^2 = 4\lambda_1^{-1}|f|^2\} = Y \cap P_nH, \qquad n = 2m; \tag{16.16}$$

i.e., $n =$ even. Then the interior normal $\nu = \nu_\Gamma(u)$ to Γ at $u \in \Gamma$ is given by $\nu = -Au/|Au|$. The coercivity condition (III) in Chapter 9 is replaced by

$$\begin{aligned}(N(u), Au) = |Au|^2 - (f, Au) &\geq \tfrac{1}{2}|Au|^2 - \tfrac{1}{2}|f|^2 \\ &\geq \tfrac{1}{2}\lambda_1|A^{1/2}u|^2 - \tfrac{1}{2}|f|^2 \geq \tfrac{3}{2}|f|^2.\end{aligned}$$

Condition (V) follows from the following fact valid for any $\xi \in T_u(\Gamma)$ (the tangent hyperplane to Γ in P_nH), $u \in \Gamma$:

$$\begin{aligned}\frac{|(I - P_n)(N(u) + \xi)|}{|P_n(N(u) + \xi)|} &\leq \frac{|(I - P_n)f||Au|}{(N(u), Au)} \leq \frac{|(I - P_n)f||Au|}{|Au|^2 - (f, Au)} \\ &\leq \frac{|(I - P_n)f||Au|}{|Au|^2 - |f||Au|} \leq \frac{|(I - P_n)f||Au|}{|Au|^2 - \frac{1}{2}|Au|^2} \\ \frac{2|(I - P_n)f|}{|Au|} &\leq \frac{|(I - P_n)f|}{|f|} \leq \frac{1}{\lambda_{n+1}}\frac{|Af|}{|f|}\end{aligned} \tag{16.17}$$

by the assumption $f \in \mathscr{D}(A)$. It is useful at this stage to introduce the quantity (spectral shape factor of f, always ≥ 1)

$$s = \frac{|Af|}{\lambda_1|f|}. \tag{16.18}$$

From (16.17) we now infer that condition (V) in Chapter 9 is satisfied provided

$$2s \leq (n + 2)\gamma. \tag{16.19}$$

As in Chapter 9, we can now use (16.17) to show that conditions (I), (II) also hold provided n is sufficiently large. However, we will prefer to check the adequate relation (9.7), which automatically will imply (I), (II); this relation is

$$\operatorname{Tr} AP(u) - \operatorname{Tr} AP_n < \frac{\lambda_{n+1} - \lambda_n}{2} \quad \text{for } u \in \Gamma, \tag{16.20}$$

where $P(u)$ denotes the orthogonal projector onto $\mathbb{R}N(u) + T_u(\Gamma)$. Proceeding as in Chapter 9 with

$$w_0 = \frac{(I - P_{T_u(\Gamma)})N(u)}{|(I - P_{T_u(\Gamma)})N(u)|} \quad \text{and} \quad v(u) = -\frac{Au}{|Au|}$$

we obtain that

$$\begin{aligned}\operatorname{Tr} AP(u) - \operatorname{Tr} AP_n &= \frac{|(I - P_n)A^{1/2}f|^2|Au|^2 - |(I - P_n)f|^2|A^{3/2}u|^2}{|(I - P_n)f|^2|Au|^2 + (|Au|^2 - (f, Au))^2} \\ &\le \frac{|(I - P_n)A^{1/2}f|^2|Au|^2}{(|Au|^2 - (f, Au))^2} \le 4\frac{|(I - P_n)A^{1/2}f|^2}{|Au|^2} \\ &\le 4\frac{|Af|^2}{\lambda_{n+1}|Au|^2} \le \frac{\lambda_1^2}{\lambda_{n+1}} s^2,\end{aligned}$$

where we used $u \in \Gamma$ and $|Au|^2 \ge \lambda_1 |A^{1/2}u|^2 = 4|f|^2$. It follows that (16.20) holds provided

$$2s^2\lambda_1^2 < \lambda_{n+1}(\lambda_{n+1} - \lambda_n),$$

that is,

$$(n + 1)(n + 2)^2 > 8s^2. \tag{16.21}$$

So in what concerns Chapter 9, it remains only to determine the constraint on n for condition (IV) to hold. To this end we notice first that for $u \in \Gamma$, $u_0 \in X$,

$$\begin{aligned}|f|/\lambda_1^{1/2} &\le |A^{1/2}u| - |A^{1/2}u_0| \le |P_n A^{1/2}u| - |P_n A^{1/2}u_0| \\ &\le |P_n(A^{1/2}u - A^{1/2}u_0)| \le \lambda_n^{1/2}|P_n(u - u_0)| \\ &= \lambda_1^{1/2}\frac{n}{2}|P_n(u - u_0)|\end{aligned} \tag{16.22}$$

and that we have for $t > 0$

$$\frac{d}{du}v + Av + B(u, v) = Af, \quad \text{where } v = Au,$$

and thus, using (16.3),

$$|Au(t)|^2 \le |Au(0)|^2 e^{-\lambda_1 t} + \frac{|Af|^2}{\lambda_1^2}(1 - e^{\lambda_1 t})$$

which implies that

$$|Au_0| \le \frac{|Af|}{\lambda_1} \quad \text{for } u_0 \in X. \tag{16.23}$$

Therefore

$$|(I - P_n)(u_0 - u)| = |(I - P_n)u_0| \le \frac{1}{\lambda_{n+1}}|Au_0| \le \left(\frac{n}{2} + 1\right)^{-2}\frac{|Af|}{\lambda_1^2}. \tag{16.24}$$

Comparing (16.22), (16.24) we obtain the desired relation

$$|(I - P_n)(u_0 - u)| \leq \tfrac{1}{3}|P_n(u_0 - u)|$$

provided

$$(n + 2)^2 \geq 6ns. \tag{16.25}$$

Noting that (16.25) implies (16.21), we can now state the following:

Lemma 16.3. *Let* Γ *be given by* (16.16) *and let* s *be given by* (16.18). *Then, if*

$$(n + 2)\gamma \geq 2s, \quad (cf.\ (16.29)), \qquad (n + 2)^2 \geq 6ns, \tag{16.26}$$

then all the properties (I) *to* (V) *in Proposition* 9.2 *hold for* Γ.

Passing now to the subject of Chapter 10, we observe the following facts. By virtue of Lemma 16.3, condition (C1) holds provided n satisfies $n + 2 \geq 6s$, $(n + 2)^2 \geq 6ns$, in particular if

$$n + 2 \geq 6s. \tag{16.27}$$

Condition (C2) holds with $\theta = 1$ (see definition (16.11) and the subsequent discussion of it). Condition (C3) is satisfied provided $n = 2m$ satisfies (16.12a), that is,

$$n = 2 \geq 40r/3. \tag{16.28}$$

Condition (C4) is satisfied for $n = 2m$, $m' = m$, if (16.7) and (16.10) hold, that is, if $n + 2 \geq 8r\sqrt{2}$ and $n^2 + n + 2 \geq 16r$, respectively; these, in turn, follow from (16.28). Finally, condition (C5) holds if $n > 2(3r)^{1/2} + 2$, which also follows from (16.28). Thus, strengthening and simplifying (16.27) and (16.28) a little bit (recall that r, $s \geq 1$), we can now state the following obvious corollary of Theorem 10.1.

Theorem 16.4. *Let* r, s, *and* Γ *be defined by* (16.13), (16.18), *and* (16.16), *respectively. Moreover, let*

$$n > 4s \quad \text{and} \quad n > 12r. \tag{16.29}$$

Then the integral surface Σ *of* (16.1), (16.1a) *constructed with* Γ *as initial data and its closure* $\bar{\Sigma}$ *enjoy all the properties given in Theorem* 10.1; *in particular,* $\bar{\Sigma}$ *is an inertial manifold of* (16.1), (16.1a).

In view of the fact that one can take $k_{14} = 2\lambda_1 r$, $k_9 = k_9' = 0$, one can prove that $\bar{\Sigma}$ is asymptotically complete provided $n \geq cr$ for a suitable large absolute constant c.

The remaining part of this section will be devoted to the study of (16.1) when $\omega \in \bigcup_{N=1}^{\infty} P_N H$, $\omega \neq 0$. We notice that these ω's are dense in H, and therefore we shall deal with an "almost generic" form of (16.1). We shall show that for this form our geometric method yields an inertial (possibly non-

Lipschitz) manifold (enjoying the asymptotic completeness property) with a dimension independent of the parameters r and s in Theorem 16.4. This class of equations is relevant for the discussion of "slaved" modes in [Ha].

We start by assuming that in (16.1), $P_N\omega = \omega$ for some $N < \infty$. Then denoting $p = P_N u$, $q = (I - P_N)u$, (16.1) becomes

$$\dot{p} - p'' + (p, \omega)p' = P_N f, \qquad p(0) = p_0 = P_N u_0, \tag{16.30}$$

$$\dot{q} - q'' + (p, \omega)q' = (I - P_N)f, \qquad q(0) = q_0 = (I - P_N)u_0. \tag{16.31}$$

One notices that (16.30) is actually an autonomous N-dimensional ordinary differential system. If u_0 and u_1 are two initial data for (16.1) and if $P_N u_0 = P_N u_1$ it follows from the uniqueness of solutions to (16.30) that $P_N(S(t)u_0) = P_N(S(t)u_1)$ for all $t \geq 0$. Denoting $p(t) = P_N S(t)u_0 = P_N S(t)u_1$, one remarks that the difference $\delta(t) = (I - P_N)S(t)u_0 - (I - P_N)S(t)u_1$ satisfies

$$\dot{\delta} - \delta'' + (p, \omega)\delta' = 0 \tag{16.32}$$

and thus

$$|\delta(t)| \leq e^{-\lambda_1 t}|\delta(0)|, \qquad t \geq 0. \tag{16.33}$$

Now let X denote the universal attractor of (16.1) and X_N that of (16.30). (Note that by our definition (see the beginning of Chapter 1), X_N may not be a small set in $P_N H$.) We first need the following.

Lemma 16.5. *The projection P_N on X is one-to-one and onto X_N.*

PROOF. Let u_0, u_1 belong to X, $P_N u_0 = P_N u_1$, and let $t_0 > 0$. Then there exist $v_0, v_1 \in X$ such that $S(t_0)v_0 = u_0$, $S(t_0)v_1 = u_1$. Then $P_N S(t)v_0 = P_N S(t)v_0$ for all $t \geq 0$ and (16.33) implies

$$|u_0 - u_1| \leq e^{-\lambda_1 t}|v_0 - v_1| \leq e^{-\lambda_1 t}\operatorname{diam}(X).$$

Letting $t \to \infty$ we obtain $u_0 = u_1$; that is, P_N is injective on X. The inclusion

$$P_N X \subset X_N$$

is obvious since if $u_0 \in X$ then $u_0 \in S(t)X$ for all $t \geq 0$ and thus $Pu_0 = S_N(t)P_N X$ also for all $t > 0$ and $P_N X$ is bounded. (Here as well as in the sequel $(S_N(t))_{t\geq 0}$ denotes the forward flow for (16.30).) It remains to prove that

$$X_N \subset P_N X. \tag{16.34}$$

To this end, let $p_0 \in X_N$, $t \geq 0$, and $p_t \in X$ such that $S_N(t)p_t = p_0$. Then $y_t = S(t)p_t$ enjoys the properties

$$y_t \in S(t)Y \subset Y \text{ (see (16.11)) } \quad \text{and} \quad P_N y_t = p_0.$$

Since Y is compact in H there exists y_∞ and y_{t_j}, $t_1 < t_2 < \cdots \to \infty$ such that $y_{t_j} \to y_\infty$. Obviously $y_\infty \in S(t)Y$ for all $t \geq 0$, thus $y_\infty \in X$ and $P_N y_\infty = \lim_{j\to\infty} P_N y_{t_j} = p_0$. This establishes (13.34) and concludes the proof of the lemma. $\square$

Let us consider $\Gamma = \{p \mid P_N p = p, |p| = R\}$ where $R = 2|f|/\lambda_1$. Set

$$\Sigma = \bigcup_{t>0} S(t)\Gamma \tag{16.35}$$

and consider the map $P_N : \Sigma \to P_N H$. We claim it is injective. Indeed, suppose $P_N u_0 = P_N u_1$, $u_0 = S(t_0)v_0$, $u_1 = S(t_1)v_1$, $v_0, v_1 \in \Gamma$. We can assume without loss of generality that $t_0 \geq t_1$. Thus

$$S_N(t_1)S_N(t_0 - t_1)P_N v_0 = P_N S(t_1)S(t_0 - t_1)v_0 = P_N S(t_1)v_1 = S_N(t_1)P_N v_1.$$

From the uniqueness property of the initial value problem for (16.30) it follows that $S_N(t_0 - t_1)P_N v_0 = P_N v_1 = v_1$. If $t_0 = t_1$ then $v_0 = v_1$, hence $u_0 = u_1$. If $t_0 - t_1 > 0$ we arrive at a contradiction since by (16.2)

$$R = |v_1| = |S_N(t_0 - t_1)v_0| = |S(t_0 - t_1)v_0| < |v_0| = R.$$

We now define

$$\Xi = X \cup \Sigma \cup \Gamma. \tag{16.36}$$

From $\Sigma \cap X = \emptyset$ and

$$P_N(\Sigma \cup \Gamma) = \bigcup_{t \geq 0} S_N(t)\Gamma = \{p \mid p \in P_N H, |p| \leq R\} \backslash X,$$

we deduce from Lemma 16.5 and the above discussion that P_N is a one-to-one map from Ξ onto $\{p \mid p \in P_N H, |p| \leq R\}$. It is also easy to see that Ξ is closed in H. Therefore the inverse map Φ of P_N restricted to Ξ is a continuous map of $\{p \mid p \in P_N H, |p| \leq R\}$. This establishes that Ξ is a (possibly nonsmooth but topological) manifold of dimension N. By its very definition Ξ is invariant, i.e., $S(t)\Xi \subset \Xi$ for all $t \geq 0$, and contains the universal attractor. We can now state the following:

Theorem 16.6. *Let $\omega \in P_N H$. Then the manifold Ξ defined by* (16.36) *is an inertial (not necessarily smooth) manifold of dimension N for* (16.1); *that is, Ξ is invariant and attracts exponentially all bounded sets. Moreover, Ξ has the asymptotic completeness property.*

Proof. Let Z be any bounded set in H. By (16.2) there exists a time $t_0 = t_0(Z)$ such that $|S(t)z| \leq R$ for all $t \geq t_0$, $z \in Z$. Define $\xi(z) = \Phi(P_N S(t_0)z)$. Then by virtue of (16.33) we have

$$|S(t + t_0)z - S(t)\xi(z)| \leq e^{-\lambda_1 t}|S(t_0)z - \xi(z)| \leq e^{-\lambda_1 t}2R.$$

Since $S(t)\xi(z) = \Phi(S_N(t)P_N S(t_0)z)$, $t \geq 0$, we infer at once that Ξ has the required attraction property as well as the asymptotic completeness property as stated in Chapter 12. □

It is noteworthy that the systems (16.30) have a complicated universal attractor and thus display nontrivial dynamics. For instance, if one chooses

$N = 4$,

$$\omega(x) = \frac{1}{\pi}\left(\cos\frac{2\pi}{L}x + \cos\frac{4\pi}{L}x\right), \tag{16.37}$$

$$(P_N f)(x) = g_1 \sin\frac{2\pi}{L}x + g_2 \sin\frac{4\pi}{L}x, \qquad g_1, g_2 \in \mathbb{R}, \tag{16.38}$$

and develops the solution $p(t)$ of (16.30),

$$p(t, x) = a_1(t)\cos\frac{2\pi}{L}x + b_1(t)\sin\frac{2\pi}{L}x + a_2(t)\cos\frac{4\pi}{L}x + b_2(t)\sin\frac{4\pi}{L}x \tag{16.39}$$

one obtains the following form of the system (16.30):

$$\begin{cases} \dot{a}_1 + \dfrac{4\pi^2}{L^2}a_1 + (a_1 + a_2)b_1 = 0, \\[2ex] \dot{b}_1 + \dfrac{4\pi^2}{L^2}b_1 - (a_1 + a_2)a_1 = g_1, \\[2ex] \dot{a}_2 + \dfrac{16\pi^2}{L^2}a_2 + 2(a_1 + a_2)b_2 = 0, \\[2ex] \dot{b}_2 + \dfrac{16\pi^2}{L^2}b_2 - 2(a_1 + a_2)a_2 = g_2. \end{cases} \tag{16.40}$$

This system, which contains two nonlinearly coupled Minea subsystems (see [CF]), has in general five fixed points. The fixed point $a_1 = 0, b_1 = (L^2/4\pi^2)g_1$, $a_2 = 0, b_2 = (L^2/16\pi^2)g_2$ has (for (g_1, g_2) in an unbounded open set of R^2) an unstable manifold of dimension 2. Thus the universal attractor of (16.40) has at least dimension 2 for g_1, g_2 appropriately chosen.

CHAPTER 17

Application: The Cahn–Hilliard Equation

In this chapter we consider the equation

$$\frac{\partial}{\partial t}u + \frac{\partial^4}{\partial x^4}u + \frac{\partial^2}{\partial x^2}p(u) = 0 \tag{17.1}$$

where

$$p(u) = -b_2 L^{-2}u - b_3 L^{-1}u^2 - b_4 u^3 \tag{17.2}$$

on the space $H = \{u \in L^2(0, L) : \int_0^L u\,dx = 0\}$ (the same as in Chapter 16) with the periodic boundary conditions. This means in particular that $A = (d^4/dx^4)u$ and $N(u) = Au - A^{1/2}p(u)$, for $u \in H^4(0, L)$ (= the L^2-Sobolev space of order 4) such that

$$\begin{aligned} u(0) = u(L), \qquad u'(0) = u'(L), \qquad u''(0) = u''(L), \\ u'''(0) = u'''(L). \end{aligned} \tag{17.3}$$

The assumption on the adimensional constants b_2, b_3, b_4 is that $b_2 < 0, b_4 > 0$. The case $b_2 > 0$ is trivial. Although the problem (17.1) to (17.3) does not enter directly in the general framework developed up to now, we shall use this problem to illustrate the flexibility of our general approach presented in Chapters 2 to 10.

We view the problem (17.1) to (17.3) as an abstract differential equation $(du/dt) + N(u) = 0$ where $N(u) = Au + R(u)$, $Au = d^4u/dx^4$, with domain $\mathscr{D}(A)$ formed by the $u \in H$ such that $u^{\text{iv}} = d^4u/dx^4 \in H$ and u, $u' = du/dx$, $u'' = d^2u/dx^2$, $u''' = d^3u/dx^3$ are L-periodic; also

$$R(u) = -\frac{\partial^2}{\partial x^2}[b_2 L^{-2}u + b_3 L^{-1}u^2 + b_4 u^3].$$

In this case

$$\lambda_{2k-1} = \lambda_{2k} = \Lambda_k = \left[\frac{2\pi k}{L}\right]^4, \qquad k = 1, 2, \ldots.$$

The basic property of (17.1)–(17.3) is that it has a Lyapunov functional:

$$V(u) = \frac{1}{2}|u'|^2 + \frac{b_2}{2L^2}|u|^2 + \frac{b_3}{3L}|u^{3/2}|^2 + \frac{b_4}{4}|u^2|^2; \tag{17.4}$$

indeed if $u = u(t)$ is a solution of (17.1)–(17.3) then

$$\frac{d}{dt}V(u) + |(u'' + p)'|^2 = 0 \tag{17.5}$$

(see [NSh]). Introducing

$$b_5 = |b_2| + \frac{|b_3|^2}{2b_4}, \qquad b_6 = 2\left[\frac{b_5}{b_4}\right]^{1/2}, \qquad b_6' = b_6 b_5^{1/2}, \qquad b_6'' = b_5 b_6$$

it is easy to verify that for the stationary solutions u_0 of (17.1)–(17.3) we have

$$|u| \le b_6 L^{-1/2}, \qquad |u'| \le b_6' L^{-3/2}, \qquad |u''| \le b_6'' L^{-5/2}. \tag{17.6}$$

From (17.4), (17.5) it now follows easily that

$$\begin{cases} \overline{\lim_{t\to\infty}}\, |u(t)| \le b_6 L^{-1/2}, & \overline{\lim_{t\to\infty}}\, |u'| \le b_6' L^{-3/2}, \\ \overline{\lim_{t\to\infty}}\, |u''(t)| \le b_6'' L^{-5/2} & \end{cases} \tag{17.7}$$

for any solution of (17.1)–(17.3). Here as well as later in this section b_7, b_8, b_8', ... will denote adimensional constants depending only on b_2, b_3, b_4.

It is also easy to see that in this case the universal attractor X is formed by the stationary solutions and all the heteroclinic and homoclinic solutions. So (using $|u|_{L^\infty}^2 \le |u||u'|$ for $u \in H^1$) on the universal attractor we have

$$\begin{aligned} V(u) &\le \frac{1}{2} b_6'^2 L^{-3} + \frac{|b_2| b_6^2}{2} L^{-3} + \frac{|b_3|}{3L}|u||u^2| + \frac{b_4}{4}|u^2|^2 \\ &\le \left[\frac{1}{2} b_6'^2 + \frac{|b_2| b_6^2}{2}\right] L^{-3} + \frac{|b_3|}{3L}|u|^{5/2}|u'|^{1/2} + \frac{b_4}{4}|u|^3|u'| \\ &\le b_7 L^{-3}, \end{aligned} \tag{17.7a}$$

where b_7 is a constant depending only on the constants b_1, b_2, b_3, and b_4. Therefore the first two relations in (17.6) will hold for $u \in X$ with b_6 and b_6' replaced by some convenient adimensional constants b_8 ($\ge b_6$) and b_8' ($\ge b_6'$), respectively. In order to apply Therem 8.4 to X we have to estimate from below

$$T = \sum_{j=1}^{l} |\phi_j''|^2 - b_2 L_2^{-2} \sum_{j=1}^{l} |\phi_j'|^2 - 2b_3 L^{-1} \sum_1^l \int_0^L (u\phi_j)''\phi_j\,dx$$
$$- 3b_4 \sum_1^l \int_0^L (u^2\phi_j)''\phi_j\,dx$$

where $u \in X$ and $(\phi_j)_1^l$ is an orthogonal system in H, contained in $\mathscr{D}(A)$. After some computation we arrive at

$$T \geq S_2 - b_2 L^{-2} S_1 - 2|b_3| L^{-1} |u|^{1/2} |u'|^{1/2} l^{1/2} S_2^{1/2} - 3b_4 \sum_1^l \int_0^L u'^2 \phi_j^2 \, dx \tag{17.8}$$

where we set $S_2 = \sum |\phi_j''|^2$, $S_1 = \sum |\phi_j'|^2$. But

$$\sum_1^l \int_0^L u'^2 \phi_j^2 \, dx \leq |u'|^2 \sum_1^l |\phi_j| |\phi_j'| \leq |u'|^2 l^{1/2} S_1^{1/2}$$

so that (17.8) becomes

$$\begin{aligned} T &\geq S_2 - b_2 L^{-2} S_1 - 2|b_3 (b_8 b_8')^{1/2} L^{-2} l^{1/2} S_2^{1/2} - 3b_4 b_8'^2 L^{-3} S_1^{1/2} l^{1/2} \\ &\geq \frac{1}{2} S_2 - 2b_3^2 b_8 b_8' l L^{-4} - \frac{9}{4} \frac{b_4^2 b_8'^4}{|b_2|} L^{-4} l = \frac{1}{2} S_2 - b_9 l L^{-4} \\ &\geq c_0 l^5 L^{-4} - b_9 l L^{-4} = c_0 L^{-4} l (l^4 - b_9/c_0), \end{aligned}$$

where we used that $S_2 \geq \lambda_1 + \cdots + \lambda_l \geq 2c_0 l^5 L^{-4}$ with an absolute constant $c_0 > 0$. From Theorem 8.4 we now infer readily that

$$d_H(X) \leq b_{10}, \qquad d_M(X) \leq 4b_{10}, \tag{17.9}$$

where b_{10} is the first integer $> (b_9/c_0)^{1/4}$.

It will be useful to remark now that from the above proof of (17.9) we also infer that the conclusion of Corollary 8.2 is valid provided

$$\lambda_1 + \cdots + \lambda_m > 2b_9 m L^{-4},$$

which in turn is assured by

$$m > (b_9/c_0)^{1/4}. \tag{17.9a}$$

Let us now implement the research program presented in Chapter 2 to 10.

To this purpose we consider only solutions $u = u(t)$ of (17.1)–(17.3) satisfying

$$|u| \leq \beta L^{-1/2}, \qquad |u'| \leq \beta' L^{-3/2}, \qquad |u''| \leq \beta'' L^{-5/2} \tag{17.10}$$

with some large fixed β, β', β'' to be chosen later.

We start with the spectral blocking property: The terms $L(t)g$, $L(t)^* g$ are given by

$$\begin{aligned} L(t)g &= -b_2 L^{-2} g'' - 2b_3 L^{-1} (ug)'' - 3b_4 (u^2 g)'', \\ L(t)^* g &= (-b_2 L^{-2} - 2b_3 L^{-1} u - 3b_4 u^2) g'', \end{aligned}$$

so that (3.7), (3.8) are satisfied with

$$\begin{cases} (\frac{1}{3}k_1)^{1/2} = [|b_3| \beta'' + 3b_4 (\beta'^3 \beta'')^{1/2} + 3b_4 (\beta\beta')^{1/2} \beta''] L^{-4} = \beta_1 L^{-4}, \\ (\frac{1}{3}k_2)^{1/2} = [\beta_1 + 4|b_3| (\beta'\beta'')^{1/2} + 12b_4 (\beta\beta'')^{1/2} \beta'] L^{-3} = \beta_2 L^{-3}, \\ (\frac{1}{3}k_3)^{1/2} = [|b_2| + 2|b_3| (\beta\beta')^{1/2} + 3b_4 \beta\beta'] L^{-2} = \beta_3 L^{-2}. \end{cases} \tag{17.11}$$

(Notice that in (3.8) we can actually take $k_1 = k_2 = 0$.) With these values of k_1, k_2, k_3, Theorem 3.2 and Corollary 3.3 remain valid in this case for all solutions satisfying (17.10). The interesting fact concerning this application is that (3.13) holds provided

$$m > \max(\beta_1^{1/3}, \beta_2^{1/2}, \beta_3). \tag{17.12}$$

Concerning the strong squeezing property, it is easy to check that (4.5) is $(Au, u) \geq |A^{1/2}u|^2$ (i.e., $k_4 = 1$, $C = 0$) and that (4.5b) is replaced in the present case by the following valid relation:

$$\begin{aligned} &|(-b_2 L^{-2} w - b_3 L^{-1}(u_1 + u_2)w - b_4((u_1^2 + u_1 u_2 + u_2^2)w)'', w)| \\ &\quad \leq \tfrac{1}{2}k_5'|A^{1/4}w|^2 + \tfrac{1}{2}k_7'|w|^2 \end{aligned} \tag{17.13}$$

for any two solutions u_1 and u_2 satisfying (17.10), where

$$k_5' = (2\beta_3 + \beta_4)L^{-2}, \qquad k_7' = \beta_4 L^{-4}, \qquad \beta_4 = 2|b_3|(\beta'\beta'')^{1/2} + 6b_4(\beta\beta'')^{1/2}\beta'. \tag{17.14}$$

Obviously the analog of (4.5a) is the particular case $u_1 = u_2 = u$ of (17.13). The equation for $|w|^2$ in Chapter 4 becomes

$$\frac{d}{dt}|w|^2 + \left[\frac{|A^{1/2}w|^2}{|w|^2} - \tfrac{1}{4}k_5'^2 - k_7'\right]|w|^2 \leq 0. \tag{17.15}$$

Therefore the condition (4.6) is replaced here by

$$\frac{\Lambda_m + \Lambda_{m+1}}{2} > 2(\tfrac{1}{4}k_5'^2 + k_7') = 4k_8' = 2\beta_5 L^{-4} \tag{17.16}$$

which obviously follows from

$$m > \frac{1}{2^{3/4}\pi}\beta_5^{1/4} \tag{17.17}$$

where

$$\beta_5 = \beta_4 + \tfrac{1}{4}(2\beta_3 + \beta_4)^2. \tag{17.18}$$

We can conclude as in Chapter 4 with the following:

Proposition 17.1. *Let u_1, u_2 be two solutions satisfying* (17.10). *If* (17.12) *and* (17.17) *hold, then either*

$$|(u_1 - u_2)(t)| \leq e^{-k_6' t}|(u_1 - u_2)(0)| \quad \text{for all } t \geq 0, \tag{17.19}$$

or there exists $t_0 \in (0, \infty)$ such that

$$|(I - P_{2n})(u_1 - u_2)(t)| \leq \tfrac{1}{3}|P_{2n}(u_1 - u_2)(t)| \quad \text{for all } t \geq t_0 \tag{17.20}$$

and for all $n \geq 4m$.

Proof. If the first alternative does not hold, then for some $t_0 > 0$ we have

$$\frac{|A^{1/2}w(t)|^2}{|w(t)|^2} \leq \frac{\Lambda_m + \Lambda_{m+1}}{2}, \qquad t \geq t_0.$$

For $n \geq m$, this yields

$$\Lambda_{n+1}|(I - P_{2n})w(t)|^2 \leq \frac{\Lambda_m + \Lambda_{m+1}}{2}|w(t)|^2,$$

that is,

$$|(I - P_{2n})w(t)|^2 \leq \frac{\Lambda_m + \Lambda_{m+1}}{2\Lambda_{n+1} - \Lambda_m - \Lambda_{m+1}}|P_{2n}w(t)|^2$$

for all $t \geq t_0$. For $n \geq 4m$

$$\frac{\Lambda_m + \Lambda_{m+1}}{2\Lambda_{n+1} - \Lambda_m - \Lambda_{m+1}} \leq \frac{1}{15} < \frac{1}{9}.$$

□

For further applications we need the following:

Lemma 17.2. *Let $u = u(t)$ be a solution of* (17.1)–(17.3) *such that*

$$|u(t)| \leq \beta L^{-1/2}, \qquad |u'(t)| \leq \beta' L^{-3/2} \quad \text{for all } t \geq 0, \tag{17.21}$$

and some parameters β, b'. Then

$$|u''(t)|^2 \leq |u''(0)|^2 e^{-\lambda_1 t} + (1 - e^{-\lambda_1 t})\beta_{10}^2 L^{-5} \quad \text{for } t \geq 0, \tag{17.22}$$

where β_{10} is a parameter depending only on $|b_2|$, $|b_3|$, b_4, and β, β'.

Proof. We have

$$\begin{aligned}
&\frac{1}{2}\frac{d}{dt}|u''|^2 + |u^{\text{iv}}|^2 \\
&\quad \leq -b_2 L^{-2}|u'''|^2 + |(L^{-1}b_3 u^2 + b_4 u^3)''||u^{\text{iv}}| \\
&\quad \leq |b_2|\beta'^{2/3}L^{-3}|u^{\text{iv}}|^{4/3} + (2|b_3|\beta'^{3/2} + 6b_4\beta^{1/2}\beta'^2)L^{-13/4}|u''|^{1/2}|u^{\text{iv}}| \\
&\qquad + (2|b_3|(\beta\beta')^{1/2} + 3b_4\beta\beta')L^{-2}|u''||u^{\text{iv}}| \\
&\quad \leq |b_2|\beta'^{2/3}L^{-3}|u^{\text{iv}}|^{4/3} + \beta^{1/4}(2|b_3|\beta'^{3/2} + 6b_4\beta^{1/2}\beta'^2)L^{-27/8}|u^{\text{iv}}|^{5/4} \\
&\qquad + \beta^{1/2}(2|b_3|(\beta\beta')^{1/2} + 3b_4\beta\beta')L^{-9/4}|u^{\text{iv}}|^{3/2} \\
&\quad = \beta_6 L^{-3}|u^{\text{iv}}|^{4/3} + \beta_7 L^{-27/8}|u^{\text{iv}}|^{5/4} + \beta_8 L^{-9/4}|u^{\text{iv}}|^{3/2},
\end{aligned}$$

where the definitions of the parameters β_6, β_7, β_8 are obvious. We thus obtain

$$\frac{d}{dt}|u''|^2 + \left|\frac{2\pi}{L}\right|^4 |u''|^2 \leq \frac{d}{dt}|u''|^2 + |u^{\text{iv}}|^2 \leq \beta_9 L^{-9} \tag{17.23}$$

where β_9 is of the form $c_1\beta_6^3 + c_2\beta_7^{8/3} + c_3\beta_8^4$ with c_1, c_2, c_3 some fixed numbers. So integrating (17.23) we obtain (17.22). □

Corollary 17.3. *If $n = 2m > b_{11}$ where b_{11} is an adimensional constant depending only on b_2, b_3, b_4, then P_n is injective on the universal attractor and $(P_n|X)^{-1}$ is a Lipschitz map.*

PROOF. We take $\beta = b_8$, $\beta' = b_8'$. Set $\beta'' = b_8''$ equal to the β_{10} yielded by Lemma 17.2. This means that on X we also have $|u''| \le b_8'' L^{-5/2}$. With this choice of β, β', β'' we let $b_{11}/8$ be equal to the maximum of the right-hand terms of (17.12), (17.17) and apply Proposition 17.1. □

Theorem 17.4. *For $n = 2m > b_{12}$, there exists an inertial manifold of dimension n that is the closure of the range of a function $\Phi : \{u \in P_n H : V(u) \le b_{13} L^{-3}\} \to H$ satisfying all the properties of Theorem 10.1; here b_{12}, b_{13} are adimensional constants depending only on b_2, b_3, and b_4.*

The proof of this theorem is similar to that of Theorem 10.1. The major differences consist in the choice of the absorbing set θY and of the initial data set Γ, which in this case will differ from those discussed in Chapters 5 and 9. In the remaining part of this section we will outline the adequate versions of Chapters 5 and 9 and we will check the analogs of conditions (C1) to (C5) in Chapter 10. We leave as an exercise for the reader all the other details of Chapter 10 for this case.

First let b_8' denote the constant β_{10} corresponding to $\beta = b_8$, $\beta' = b_8'$. It follows easily from Lemma 17.2 that

$$|u'| \le b_8' \quad \text{for } u \in X. \tag{17.24}$$

We recall that, by definition, b_8 and b_8' satisfy

$$|u| \le b_8 L^{-1/2}, \qquad |u'| \le b_8' L^{-3/2} \quad \text{if } V(u) \le b_7 L^{-3}. \tag{17.25}$$

Actually we have the more general property

$$\tfrac{1}{2}|u'|^2 + \frac{b_4}{16} L^{-1} |u|^4 \le V(u) + b_{14} L^{-3} \tag{17.25a}$$

where

$$b_{14} = \frac{4}{b_4}\left(2|b_2| + \frac{b_3^2}{b_4}\right)^2. \tag{17.25b}$$

Moreover, we have also the following:

Lemma 17.5. *For adequate constants b_{15}, b_{16}, $b_{17} > b_7$, we have*

$$V(u) \le b_{15} L^4 |(u'' + p)'|^2 \tag{17.26}$$

and

$$L^4 |p'|^2 \le b_{16} V(u) \tag{17.27}$$

provided $V(u) \ge b_{17} L^{-3}$.

PROOF. Setting $\gamma = L^{7/2} |(u'' + p)'|$ we have $|u'' + p| \le \gamma / 2\pi L^{5/2}$ and, by considering $(u'' + p, u)$,

$$|u'|^2 + b_2 L^{-2} |u|^2 + b_3 L^{-1} |u^{3/2}|^2 + b_4 |u^2|^2 \le |u| \gamma / 2\pi L^{5/2},$$

whence

$$|u'|^2 + \tfrac{1}{8}b_4 L^{-1}|u|^4 \le \left[2b_5 b_4^{-1} + \tfrac{3}{4}b_4^{-1/3}\left(\frac{\gamma}{2\pi}\right)^{4/3}\right]L^{-3}. \tag{17.28}$$

On the other hand, $V(u)$ has the following upper estimate:

$$V(u) \le \tfrac{1}{2}|u'|^2 + b_{18}|u'||u|^3 \le |u'|^2 + \frac{b_{18}^2}{2}|u|^6 \tag{17.29}$$

where

$$b_{18} = \frac{b_4}{4} + \frac{b_3^2}{18|b_2|}. \tag{17.29a}$$

It follows that either $|u'|^2 > \frac{1}{2}V(u)$ or $|u|^6 \ge V(u)/b_{18}^2$. In the first case

$$V(u)L^3 \le 2b_5 b_4^{-1} + (3/(4b_4^{1/3}(2\pi)^{2/3}))\gamma^{4/3}, \tag{17.30}$$

while in the second case

$$(L^3 V(u))^{2/3} \le 16b_{18}^{1/3} b_5 b_4^{-2} + (6b_{18}^{1/3}/(2\pi b_4)^{4/3})\gamma^{4/3}. \tag{17.31}$$

We take

$$b_{17} = \max\{4b_5 b_4^{-1}, (3/(2b_{17}^{1/3} b_5 b_4^{-2}))^{3/2}, 2b_7, 2, b_{14}\}, \tag{17.32}$$

$$b_{15} = \max\{(3/2b_4^{1/3}(2\pi)^{4/3})^{3/2}, (12b_{18}^{1/3}/(2\pi b_4)^{4/3})^{3/2}\}, \tag{17.32a}$$

and then from (17.30) and (17.31), we infer easily that

$$(L^3 V(u))^{2/3} \le \frac{b_{17}^{2/3}}{2} + \frac{b_{15}^{2/3}}{2}\gamma^{4/3}.$$

Therefore if $L^3 V(u) \ge b_{17}$ we obtain (17.26). In this case we also obtain (cf. (17.25a))

$$|u'|^2 \quad \text{and} \quad L^{-1}|u|^4 \le b_{19} V(u). \tag{17.33}$$

The relation (17.27) now follows easily from (17.33). Indeed

$$\begin{aligned}
L^{7/2}|p'| &\le L^{3/2}|b_2||u'| + 2L^{5/2}|b_3||uu'| + 3b_4 L^{7/2}|u^2u'| \\
&\le L^{3/2}|b_2||u'| + 2L^{5/2}|b_3||u|^{1/2}|u'|^{3/2} + 3b_4 L^{7/2}|u||u'|^2 \\
&\le b_{20}(L^3 V(u))^{1/2} + b_{20}'(L^3 V(u))^{7/8} + b_{20}''(L^3 V(u))^{5/4} \le b_{16}^{1/2}(L^3 V(u))^{5/4}
\end{aligned}$$

where $b_{20}, b_{20}', b_{20}''$ are adequate constants and

$$b_{16} = (b_{20} b_{14}^{1/2} + b_{20}' b_{14}^{7/8} + b_{20}'' b_{14}^{5/4})^2 |b_{14}^{5/2}|. \tag{17.34}$$ □

We can now define the absorbing sets analogous to θY ($1 \le \theta < \infty$). We first define b_{21} to be the parameter β_{10} given by Lemma 17.2 corresponding to the choices $\beta = (4b_{17}b_{19})^{1/2}$, $\beta' = \beta^2$. Now for $\alpha \ge 0$ we define

$$Y_\alpha = \{u \in D(A^{1/2}) : V(u) \le 4b_{17}L^{-3}, |A^{1/2}u| \le (b_{21} + 4b_{17}b_{19})L^{-5/2}\}. \tag{17.35}$$

Lemma 17.6. *Y is a convex compact set in H and a neighborhood of 0 in $\mathscr{D}(A^{1/2})$. Moreover, Y is absorbing (in H).*

PROOF. We prove only the last statement since the other ones are obvious. We notice first that if $V(u_0) \geq b_{17}L^{-3}$ and $|u_0''| \geq (b_{21} + 4b_{17}b_{19})L^{-5/2}$, then by virtue of Lemmas 17.2 and 17.5 we have (for $u(t) = S(t)u_0$ and $t \geq 0$)

$$|u''(t)|^2 \leq |u_0''|^2 e^{-\lambda_1 t} + (1 - e^{-\lambda_1 t})b_{21}^2 L^{-5} \leq |u_0''|^2 L^{-5} \tag{17.36}$$

and

$$\frac{d}{dt} V(u(t)) + b_{15}^{-1} L^{-4} V(u(t)) \leq 0;$$

hence

$$V(u(t)) \leq V(u_0) e^{-b_{15}^{-1} L^{-4} t}. \tag{17.36a}$$

In particular, we easily deduce

$$S(t) Y_\alpha \subset Y_\alpha, \qquad t \geq 0. \tag{17.37}$$

On the other hand, for $u = S(t)u_0$, $|u_0| L^{1/2} \geq (b_{17}b_{19})^{1/2}$ we also have

$$\begin{aligned}
\frac{1}{2}\frac{d}{dt}|u|^2 + |A^{1/2}u|^2 &\leq |A^{1/2}u| \left(\frac{|b_2|}{L^2}|u| + \frac{|b_3|}{L}|u|^2 + b_4 |u^3| \right) \\
&\leq |A^{1/2}u| \left(\frac{|b_2|}{L^2}|u| + \frac{|b_3|}{L}|u'|^{1/2}|u|^{3/2} + b_4 |u|^2 |u'| \right) \\
&\leq |A^{1/2}u| \frac{|b_2|}{L^2}|u| + |A^{1/2}u|^{5/4} \frac{|b_3|}{L}|u|^{7/4} + |A^{1/2}u|^{3/2} b_4 |u|^{5/2} \\
&\leq \frac{1}{2}|A^{1/2}u|^2 + b_{22}\left(\frac{1}{L^4}|u|^2 + |u|^{10} \right).
\end{aligned}$$

It follows easily that there exist constants b_{23}, b_{24} such that for

$$0 \leq t \leq t_0 = b_{23} L^4 (L^{1/2}|u_0|)^{-8} \tag{17.38}$$

we have $|u(t)| \leq 2|u_0|$ and

$$\frac{1}{t_0} \int_0^{t_0} |A^{1/2}u|^2 \, dt \leq b_{24} L^{-5} (L^{1/2}|u_0|)^{10}. \tag{17.38a}$$

After some computations we obtain

$$\frac{1}{t_0} \int_0^{t_0} V(u) \, dt \leq b_{25} L^{-3} (L^{1/2}|u_0|)^6. \tag{17.38b}$$

By (17.38a, b), there exist $t_1, t_2 \in (0, t_0)$ such that

$$\begin{aligned}
|A^{1/2}u(t_1)|^2 &\leq b_{24} L^{-5} (L^{1/2}|u_0|)^{10}, \\
V(u(t_2)) &\leq b_{25} L^{-3} (L^{1/2}|u_0|)^6.
\end{aligned}$$

The discussion at the beginning of the proof shows that

$$|u''(t_0)|^2 = (|A^{1/2}u(t_0)|^2) \leq L^{-5}\max\{b_{21}^2, b_{24}(L^{12}|u_0|)^{10}\},$$
$$V(u(t_0)) \leq L^{-3}\max\{b_{17}, b_{25}(L^{1/2}|u_0|)^6\}. \quad (17.39)$$

The fact that Y_α is absorbing now follows easily from (17.38), (17.39) and (17.36), (17.36a). □

It is clear that Y_α satisfies conditions analogous to (5.1), (5.2), and (5.3). We now define

$$\Gamma = \{u = P_n H : V(u) = 4\beta_{11} b_{17} L^{-3}\} \quad (17.40)$$

where $n = 2m$ is even and $\beta_{11} \geq 1$ is a parameter to be chosen later. Notice that if $u \in \Gamma$ then, by virtue of (17.33),

$$|u'| \leq (4b_{19}b_{17}\beta_{11})^{1/2}L^{-3/2}$$

and therefore

$$|u''| = |A^{1/4}u'| \leq \frac{\pi n}{L}|u'| \leq (4b_{19}b_{17}\beta_{11})^{1/2}L^{-5/2}\pi n.$$

We conclude that

$$\Gamma \subset \beta_{11} Y_{\pi n}. \quad (17.41)$$

This is the analog of condition (C2) in Chapter 10. The analog of condition (C5) is in this case a relation similar to (17.9a), i.e.,

$$n > b_{26}. \quad (17.42)$$

By (17.12), condition (C4) becomes

$$m > \max\{\beta_1^{1/3}, \beta_2^{1/2}, \beta_3\} \quad (17.43)$$

where $\beta_1, \beta_2, \beta_3$ are computed according to (17.11) and (see also (17.33))

$$\beta = (4\beta_{11}b_{17}b_{19})^{1/4}, \qquad \beta' = (4\beta_{11}b_{17}b_{19})^{1/2}, \qquad \beta'' = \beta'\pi n. \quad (17.44)$$

It is easy to check that

$$\beta_1, \beta_2 \leq b_{27}\beta_{11}n^{1/2} + b_{28}\beta_{11}^{7/8}n, \qquad \beta_3 \leq b_{29}\beta_{11}^{3/4} \quad (17.44a)$$

so that (17.43) is ensured provided

$$n \geq b_{30}\beta_{11}^{7/8}. \quad (17.45)$$

Our next step is to find the conditions under which Theorems 5.1 and 5.2 remain valid in $\beta_{11} Y_{\pi m}$. In other words, we will now check the analog of condition (C3) in Chapter 10. For $u_1^0, u_2^0 \in \beta_{11} Y_{\pi n}$, $w \in \mathscr{D}(A)$, we have

$$\tilde{L}(t)v = -b_2 L^{-2}w'' - b_3 L^{-1}((u_1 + u_2))w)'' - b_4((u_1^2 + u_1 u_2 + u_2^2)w)''$$

where $u_j = S(t)u_i^0$, $i = 1, 2$, and $t \geq 0$. It follows that (5.5) holds with

$$k_9 = L^{-2}b_{31}\beta_{11}^{3/4}, \qquad k_9' = L^{-4}b_{32}\beta_{11}^{5/4}$$

Therefore relation (5.6) (with $\alpha = \frac{1}{3}$) becomes

$$\left(\frac{2\pi(m+1)}{L}\right)^4 - \left(\frac{2\pi m}{L}\right)^4$$
$$\geq \frac{b_{31}\beta_{11}^{3/4}}{L^2}\left(\frac{2\pi m}{L}\right)^2 + \frac{13}{3}(3\beta_1^2 + b_{32}^2\beta_{11}^{5/2} + 3\beta_2^2(2\pi m)^2 + 3\beta_3^2(2\pi m^4))^{1/2}\frac{1}{L^4}, \tag{17.46}$$

where we used (17.11) and $n = 2m$. Taking into account (17.44a), we now obtain, after some computations, that (17.46) holds provided

$$n \geq b_{33}\beta_{11}^{7/8}. \tag{17.47}$$

We will replace b_{30} in (17.45) by $\max\{b_{26}, b_{30}, b_{33}\}$ so that (17.45) will also imply (17.47), (17.42).

Concluding, we have proved that conditions (C2), (C3), (C4), (C5) will hold in our present case provided Γ is defined by (17.40), θY is replaced by $\beta_{11} Y_{\pi n}$, and (17.45) holds.

Our final step consists in proving the requirements of Γ given in Chapter 9. The role of the coercivity condition (III) in Chapters 9 and 10 is played by (17.26) in the present case, so we will concentrate on the other properties (I) to (V) in Chapter 9.

We start by discussing property (IV). Let $u \in \Gamma$, $u_0 \in X$. First we notice

$$\begin{aligned}|(1 - P_n)(u - u_0)| &\leq \Lambda_{m+1}^{-1/4}|A^{1/4}(I - P_n)(u - u_0)| \\ &\leq \frac{L}{2\pi(m+1)}|A^{1/4}(u - u_0)| = \frac{L}{2\pi(m+1)}|u' - u_0'| \\ &\leq \frac{1}{nL^{1/2}} b_{34}\beta_{11}^{1/2}\end{aligned}$$

and second

$$\begin{aligned}|P_n(u - u_0)| &\geq |u - u_0| - |(I - P_n)(u - u_0)| \\ &\geq \beta_{11}^{1/4}(b_{17}b_{19})^{1/4}L^{-1/2} - b_8 L^{-1/2} - \frac{1}{nL^{1/2}}b_{34}\beta_{11}^{1/2} \\ &\geq \frac{3}{nL^{1/2}} b_{34}\beta_{11}^{1/2}.\end{aligned} \tag{17.48}$$

It is obvious that by setting $b_{37} = \max\{1, b_{35}\}$ and replacing b_{30} by $\max\{b_{30}, b_{36}\}$ we ensure that

$$\beta_{11} \geq b_{37} \tag{17.49}$$

and (17.45) imply property (IV) for Γ. So it remains to verify that properties (I), (II), and (V) in Chapter 9 can also be satisfied by our present Γ. To this purpose we shall present here a method which differs from that presented in Chapter 9. The emphasis here lies on the following:

Lemma 17.7. *If*

$$n \geq b_{38}\beta_{11}^{3/2} \tag{17.50}$$

then all $u \in \Gamma$, $\xi \in T_u(\Gamma)$ satisfy

$$\rho = \frac{|(I - P_n)A^{1/2}(N(u) + \xi)|}{|(N(u), v)|} \leq b_{39}\beta_{11}^{3/2}\frac{n}{L^2} \tag{17.51}$$

where v is the normal (in $P_n(H)$) to Γ at u and b_{38}, b_{39} are as usual some adequate constants depending only on b_2, b_3, b_4.

Proof. The normal (in P_nH) to Γ and u is

$$v = -\frac{P_n(u'' + p)}{|P_n(u'' + p)|};$$

hence

$$\begin{aligned}(N(u), v) &= \frac{|P_n(u'' + p)'|^2}{|P_n(u'' + p)|} \geq \frac{2\pi}{L}|P_n(u''' + p')| \\ &\geq \frac{2\pi}{L}(|u'''| - |p'|) \geq \frac{2\pi}{L}[|u'''| - 2(b_{16}b_{17}\beta_{11}L^{-7})^{1/2}]\end{aligned} \tag{17.52}$$

by virtue of (17.27). On the other hand,

$$\begin{aligned}|(I - P_n)A^{1/2}(N(u) + \xi)| &= |(I - P_n)p^{\text{iv}}| \\ &= \left|(I - P_n)\left(\frac{b_3}{L}u^2 + b_4u^3\right)^{\text{iv}}\right| \\ &\leq \left|\left(\frac{b_3}{L}u^2 + b_4u^3\right)^{\text{iv}}\right| \\ &\leq \frac{|b_3|}{L}|(u^2)^{\text{iv}}| + b_4|(u^3)^{\text{iv}}|.\end{aligned} \tag{17.53}$$

After some easy but tedious computations one obtains

$$\begin{cases}|(u^2)^{\text{iv}}| \leq c_1|u|^{1/2}|u'|^{1/2}|u^{\text{iv}}|, \\ |(u^3)^{\text{iv}}| \leq c_2|u||u'||u^{\text{iv}}|,\end{cases} \tag{17.54}$$

where c_1, c_2 are some (absolute) constants. Introducing (17.54) in (17.53) and using (17.33) and $P_nu = u$, we obtain

$$\begin{aligned}|(I - P_n)A^{1/2}(N(u) + \xi)| &\leq b_{40}L^{-3}(\beta_{11}^{3/8} + b_{11}^{3/2})|u'''|n \\ &\leq 2b_{40}L^{-3}\beta_{11}^{3/2}n|u'''|.\end{aligned} \tag{17.55}$$

Now, if

$$|u'''| > 4(b_{16}b_{17}\beta_{11}L^{-7})^{1/2}$$

then (17.52) and (17.55) yield

$$\rho \le (b_{40}/\pi)\beta_{11}^{3/2} n/L^2. \tag{17.56}$$

If

$$|u'''| \le 4(b_{16}b_{17}\beta_{11}L^{-7})^{1/2} \tag{17.57}$$

then on the one hand

$$|(I - P_n)A^{1/2}(N(u) + \xi)| \le b_{41}L^{-13/2}\beta_{11}^2 n \tag{17.58}$$

and on the other hand

$$\begin{aligned}
(N(u), v) &\ge \frac{2\pi}{L}|P_n(u''' + p')| \ge \frac{2\pi}{L}(|(u''' + p')| - |(I - P_n)p'|) \\
&\ge \frac{2\pi}{L}2(b_{15}^{-2}b_{17}\beta_{11}L^{-7})^{1/2} - \frac{L^2}{(2\pi)^2 m^3}|(I - P_n)|\frac{b_3}{L}|(u^2 + b_4u^3)^{\mathrm{iv}}| \\
&\ge L^{-9/2}b_{42}\beta_{11}^{1/2} - \frac{2L^2}{\pi^2 n^3}2b_{40}L^{-3}\beta_{11}^{3/2}n|u'''| \\
&\ge (L^{-9/2}b_{42}\beta_{11}^{1/2} - L^{-1}b_{43}\beta_{11}^{3/2}n^{-2}|u'''|) \\
&\ge L^{-9/2}(b_{42}\beta_{11}^{1/2} - b_{44}\beta_{11}^2 n^{-2}),
\end{aligned}$$

where we used (17.26), the estimates leading to (17.55), and (17.57). Hence if

$$n > \frac{2b_{44}}{b_{42}}\beta_{11}^{3/2} \tag{17.59}$$

then

$$(N(u), v) \ge L^{-9/2}b_{42}\beta_{11}^{1/2}/2;$$

this and (17.58) yield

$$\rho \le (2b_{41}/b_{42})\beta_{11}^{3/2}n/L^2.$$

The proof is concluded by setting

$$b_{38} = 2b_{44}/b_{42}, \qquad b_{39} = \max\{b_{40}/\pi, 2b_{41}/b_{42}\}. \qquad \square$$

An obvious consequence of Lemma 17.7 is the following.

Corollary 17.8. *If* (17.50) *holds, then*

$$\frac{|(I - P_n)(N(u) + \xi)|}{|(N(u), v)|} \le \frac{b_{39}\beta_{11}^{3/2}}{2\pi^2 n}. \tag{17.60}$$

We can now find the supplementary conditions on n and β_{11} in order that property (V) in Chapter 9 be satisfied by our present Γ. Indeed, if

$$n > 3b_{39}\beta_{11}^{3/2}/2\pi^2 \tag{17.61}$$

and if (17.50) is also valid, then by virtue of (17.60), property (V) will hold (with $\gamma = \frac{1}{3}$). Now setting

$$b_{45} = \max\{3b_{39}/2\pi^2, b_{38}, b_{30}\}$$

we ensure that if β_{11} satisfies (17.49) and if

$$n > b_{45}\beta_{11}^{3/2} \tag{17.62}$$

then properties (III), (IV), and (V) all hold for Γ. In order to check the only remaining properties, (I) and (II), we will now prove Lemma 17.9.

Lemma 17.9. *Let n and β_{11} satisfy* (17.62) *and* (17.49); *then for all $u \in \Gamma$ we have*

$$\lambda(u) \geq \Lambda_{m+1}\left(1 - \frac{b_{39}^2\beta_{11}^3}{4\pi^4 n^2}\right), \tag{17.63}$$

$$\Lambda(u) \leq \Lambda_m\left(1 + \frac{b_{39}^2\beta_{11}^3}{4\pi^4 n^2}\right), \tag{17.64}$$

where $\lambda(u)$ and $\Lambda(u)$ are defined as in Chapter 9 (*in the paragraph introducing properties* (I) *and* (II)).

Proof. Let $u \in \Lambda$ and $e \in \mathscr{D}(A)$, $P(u)e = 0$. Then $e = \alpha v + f$ where v was introduced in Lemma 17.7, $\alpha \in \mathbb{R}$, and $f \in \mathscr{D}(A)$, $P_n f = 0$. Since e is also orthogonal on $N(u)$ we have

$$\alpha = -\frac{(f, (I - P_n)N(u))}{(N(u), v)};$$

hence, by Corollary 17.8,

$$|\alpha| \leq |f|\frac{b_{39}\beta_{11}^{3/2}}{2\pi^2 n^2}.$$

It follows that

$$|f|^2\left(1 + \frac{b_{31}^2\beta_{11}^3}{4\pi^4 n^2}\right) > |f|^2 + \alpha^2 = |e|^2 \tag{17.65}$$

and

$$(Ae, e) \geq (Af, f) \geq \Lambda_{m+1}|f|^2. \tag{17.66}$$

Introducing (17.65) into (17.66), we readily obtain (17.63). As for (17.64), we take $g = N(u) + \xi$ with $u \in \Gamma$, $\xi \in T_u(\Gamma)$ and notice

$$\begin{aligned}(Ag, g) = |A^{1/2}g|^2 &\leq |(I - P_n)A^{1/2}(N(u) + \xi)|^2 + |A^{1/2}P_n(N(u) + \xi)|^2 \\ &\leq b_{39}^2\beta_{11}^3\frac{n^2}{L^4}(N(u), v)^2 + \Lambda_m|P_n(N(u) + \xi)|^2,\end{aligned}$$

where we used (17.51). But

$$|P_n(N(u) + \xi)|^2 = (N(u), v)^2 + |P_n N(u) - (N(u), v)v + \xi|^2$$

so that

$$(Ag, g) \leq \left(b_{39}^2 \beta_{11}^3 \frac{n^2}{L^4} + \Lambda_m \right) |P_n(N(u) + \xi)|^2 \leq \left(\frac{b_{39}^2 \beta_{11}^3}{4\pi^4 n^2} + 1 \right) \Lambda_m |P_n g|^2.$$

Obviously this establishes (17.64). □

Now if (17.62) and (17.49) are valid we have by (17.63) and (17.64)

$$\lambda(u) - \frac{\Lambda_m + \Lambda_{m+1}}{2} \leq \frac{4}{L^4} m^2 (n - 4b_{39}^2 \beta^3) > 0,$$

$$\frac{\Lambda_{m+1} + \Lambda_m}{2} - \Lambda(u) \leq \frac{4}{L^4} m^2 (n - b_{39}^2 \beta^3) > 0,$$

provided

$$n > 4b_{39}^2 \beta^3. \tag{17.67}$$

Thus (17.62), (17.67), and (17.49) also imply properties (I) and (II) for Γ. Now setting

$$b_{46} = \max\{b_{45}, 4b_{39}^2\}$$

we conclude that all conditions (C1) to (C5) in Chapter 10 are satisfied provided

$$n > b_{46} \beta_{11}^3 \tag{17.68}$$

and β_{11} satisfies (17.49). We now set $\beta_{11} = b_{37}$ (making (17.49) valid), then $b_{12} = b_{46} b_{37}^3$ and $b_{13} = 4b_{37} b_{17}$ (see (17.40)). We see then that

$$\Gamma = \{u = P_n H : V(u) = b_{13} L^{-3}\}$$

satisfies, for

$$n > b_{12},$$

all conditions (C1) to (C5) in Chapter 10. We can show then as in Chapter 10 that the closure of the integral manifold generated by Γ is an inertial manifold. This establishes Theorem 17.4.

Remark 17.10. It is easy to prove that if b_{47}/b_{12} is large enough and if $n > b_{47}$ then the inertial manifold give in Theorem 17.4 also enjoys the supplementary properties considered in Chapters 12 and 13. Finally, the approximation property considered in Chapter 14 is, of course, also true for our present inertial manifold.

CHAPTER 18

Application: A Parabolic Equation in Two Space Variables

In this section we illustrate the method developed in Chapters 3 to 13 on a simple semilinear reaction–diffusion equation in two space variables.

We consider the equation

$$u_t - \Delta u = f(u) + g, \tag{18.1}$$

$$u(0) = u_0, \tag{18.2}$$

in $\Omega = [-\pi, \pi]^2$. We impose periodic boundary conditions. The function g is assumed to be smooth, periodic, and odd. The function $f(u)$ is assumed to satisfy the conditions

$$f \quad \text{odd, smooth,} \tag{18.3}$$

$$f'(u) \leq \delta < 1, \tag{18.4}$$

$$|f'(u)| \leq c_1(1 + |u|^p), \qquad |f''(u)| \leq c_1(1 + |u|^p), \qquad u \in \mathbb{R}, \tag{18.5}$$

where $0 < p < 1$. We consider H, the L^2 space of odd periodic functions in $[-\pi, \pi]^2$. We shall denote by A the self-adjoint realization of $-\Delta$ in H. The eigenvalues of A, when counted with multiplicities, are denoted by $(\lambda_j)_{j=1}^{2}$. Obviously $\lambda_1 = 1$; also it is well known that λ_j/j has a positive limit as $j \to \infty$. Equation (18.1) can be written as a differential equation in H:

$$u_t + N(u) = 0 \quad \text{with } N(u) = Au - f(u) + g. \tag{18.6}$$

We sketch here for completeness the proof of the global existence of solutions of (18.6). We note first that since $f'(u) \leq \delta$, $uf(u) \leq \delta u^2$, and $\lambda_1 = 1$ we obtain immediately from (18.6)

$$|u(t)|^2 \le |u(0)|^2 e^{-(1-\delta)t} + \frac{1}{(1-\delta)^2}|g|^2(1 - e^{-(1-\delta)t}), \qquad t \ge 0, \quad (18.7)$$

$$|A^{1/2}u(t)|^2 \le |A^{1/2}u(0)|^2 e^{-(1-\delta)t} + \frac{1}{(1-\delta)^2}|g|^2(1 - e^{-(1-\delta)t}), \qquad t \ge 0. \quad (18.8)$$

Thus if $|u(0)| \le R_0$ (resp. $|A^{1/2}u(0)| \le R_0$) for some $R_0 > |g|/(1-\delta)$, it follows that $|u(t)| \le R_0$ (resp. $|A^{1/2}u(t)| \le R_0$) for all $t \ge 0$.

For the estimate of $|Au(t)|$ we use

$$\begin{aligned} \frac{1}{2}\frac{d}{du}|Au(t)|^2 + |A^{3/2}u(t)|^2 &\le \delta|Au(t)|^2 + |A^{1/2}g||A^{3/2}u(t)| \\ &\quad + c_1 \int_\Omega (1 + |u(t)|^p)|\nabla u(t)|^2|\Delta u(t)|\,dx. \end{aligned} \quad (18.9)$$

We use the estimates for smooth and periodic functions ϕ

$$\begin{cases} |\phi|_{L^\infty} \le c_2|A^{1/2}\phi|(1 + \log|A\phi|/|A^{1/2}\phi|)^{1/2}, \\ |\phi|_{L^q} \le c_3(q)|A^{1/2}\phi| \quad \text{for all } q > 2. \end{cases} \quad (18.10)$$

The last term in (18.9) is bounded above by

$$(1 + |u|_{L^\infty}^p)|\Delta u|_{L^q}|A^{1/2}u|_{L^r}^2 \quad \left(\text{with } r = \frac{2q}{q-1}\right).$$

In its turn this is less than

$$\begin{aligned} &c_3(q)(1 + |u|_{L^\infty}^p)|A^{3/2}u||A^{1/2}u|_{L^\infty}^{2/q}|A^{1/2}u|^{4r} \\ &\le c_4(q)\left[1 + |A^{1/2}u|^p\left(1 + \log\frac{|Au|}{|A^{1/2}u|}\right)^{p/2}\right]|A^{3/2}u||A^{1/2}u|^{4r} \\ &\quad \cdot |Au|^{2/q}\left(1 + \log\frac{|A^{3/2}u|}{|Au|}\right)^{1/q}. \end{aligned}$$

Thus assuming that $|A^{1/2}u(0)| \le R_0$, $R_0 \ge \max\{1, |g|/(1-\delta)\}$ in (18.9) is bounded above by

$$\begin{aligned} &2c_4(q)R_0^{p+4r}[1 + \log(|A^{3/2}u|/|Au|)]^{(pq+2)/2q}|Au|^{2/q}|A^{3/2}u| \\ &\le 2c_5(q,\varepsilon)R_0^{p+4r}\left(\frac{|A^{3/2}u|}{|Au|}\right)^\varepsilon |Au|^{2/q}|A^{3/2}| \end{aligned}$$

for any $\varepsilon > 0$. Choosing $\varepsilon = 2/q$, we obtain finally that

$$\frac{1}{2}\frac{d}{dt}|Au|^2 + |A^{3/2}u|^2 \le \delta|Au|^2 + |A^{1/2}g||A^{3/2}u| + c_6(q)R_0^{p+4r}|A^{3/2}u|^{1+2q},$$

whence

$$\frac{d}{dt}|Au|^2 + |A^{3/2}u|^2 \leq \delta|Au|^2 + |A^{1/2}g|^2 + c_7(q)R_0^r.$$

By integrating this inequality and by using $|A^{3/2}u| \geq |Au|$ we obtain

$$|Au(t)|^2 \leq |Au(0)|^2 e^{-(1-\delta)t} + \frac{|A^{1/2}g|^2 + c_7(q)R_0^r}{1-\delta}(1 - e^{-(1-\delta)t}), \qquad t \geq 0. \tag{18.11}$$

Thus if

$$R_1 \geq \left(\frac{|A^{1/2}g|^2 + c_7(q)R_0^r}{1-\delta}\right)^{1/2}, \qquad R_0 \geq \max\left\{\frac{|g|}{1-\delta}, 1\right\}, \tag{18.12}$$

then $|A^{1/2}u(0)| \leq R_0$ and $|Au(0)| \leq R_1$ imply $|A^{1/2}u(t)| \leq R_0$ and $|Au(t)| \leq R_1$ for all $t \geq 0$. In this case we also have

$$\begin{aligned} |u(t)|_{L^\infty} &\leq c_2|A^{1/2}u(t)|\left(1 + \log\frac{|Au(t)|}{|A^{1/2}u(t)|}\right)^{1/2} \\ &\leq C_2R_0(1 + \log R_1)^{1/2}, \qquad t \geq 0. \end{aligned} \tag{18.13}$$

This is obvious if $|A^{1/2}u(t)| \geq 1$; if $|A^{1/2}u| \leq 1$ one has to study the function $\xi(1 + \log(|Au(t)|/\xi))^{1/2}$ on $(0, 1]$.

The global attractor for (18.6) is formed by the single stationary solution: $X = \{u_\infty\}$. One can easily see this by considering two solutions $S(t)u_0 = u(t)$, $S(t)v_0 = v(t)$ of (18.6) and by using

$$(N(u) - N(v), u - v) \geq (1 - \delta)|u - v|^2.$$

Indeed, one obtains

$$|u(t) - v(t)| \leq |u_0 - v_0|e^{-(1-\delta)t}, \qquad t \geq 0.$$

Since (18.6) has at least one stationary solution u_∞, the above relation shows that this solution is unique and that all solutions converge to u_∞. Thus indeed $X = \{u_\infty\}$. The estimates on the solutions of (18.6) also yield

$$\begin{aligned} &|u_\infty| \leq |g|(1-\delta)^{-1}, \qquad |A^{1/2}u_\infty| \leq |g|(1-\delta)^{-1}, \\ &|Au_\infty| \leq c_8(q)(1 + |A^{1/2}g|)^{r/2}(1-\delta)^{-(1+r)/2}, \end{aligned} \tag{18.14}$$

where $c_8(q) = 1 + (1 + c_7(q))^{1/2}$ and

$$\begin{aligned} |u_\infty|_{L^\infty} &\leq c_9(q)\frac{|A^{1/2}g|}{1-\delta}\left(1 + \log\frac{c_8(q)(1 + |A^{1/2}g|)}{(1-\delta)^{3/2}}\right)^{1/2} \\ &\leq c_{10}(q,\delta)|A^{1/2}g|(1 + \log(1 + |A^{1/2}g|))^{1/2}. \end{aligned} \tag{18.15}$$

We remark that the condition $p < 1$ was not required up to this point.

We start now checking the conditions (C1) to (C5) in Chapter 10. We shall take as initial data for our integral manifold Σ the ellipsoid

$$\Gamma = \{u \mid P_n u = u, |A^{1/2}u| = R\}$$

with R and n large enough, to be determined later. The outward normal (in P_nH) to Γ at $u \in \Gamma$ is $\nu(u) = Au$. The coercivity condition (III) in Chapter 9 is replaced here by $(\nu(u), N(u)) > 0$, $u \in \Gamma$. We note that

$$(Au, N(u)) \geq \frac{1-\delta}{2}|Au|^2 \tag{18.16}$$

if

$$R^2 \geq 4|g|^2/(1-\delta). \tag{18.16a}$$

Thus (18.16a) ensures that Γ has property (III) in Chapter 9. Concerning properties (I) and (II), an inspection of the proof of Proposition 9.2 reveals that Γ has properties (I) and (II) if

$$\frac{|Au|^2|A^{1/2}(I-P_n)N(u)|^2}{(N(u), Au)^2} < \frac{\lambda_{n+1}-\lambda_n}{2} \tag{18.17}$$

for all $u \in \Gamma$ (see relation (9.12) with u replaced by $\nu(u) = Au$). Now for $u \in \Gamma$, $(I - P_n)N(u) = -(I - P_n)(f(u) + g)$ and thus

$$|A^{1/2}(I-P_n)N(u)|^2 \leq 3\lambda_{n+1}^{-1}(|f'(u)|_{L^\infty}^2|Au|^2 + |f''(u)|_{L^\infty}^2(|\nabla u|_{L^4}^4 + |Ag|^2)).$$

Using (18.5) and (18.10) and assuming $R \geq 1$ we obtain that, for $u \in \Gamma$,

$$|f'(u)|_{L^\infty} \quad \text{and} \quad |f''(u)|_{L^\infty} \leq c_{11}R^p(1+\log\lambda_n)^{p/2};$$

also we estimate, for $u \in \Gamma$,

$$|\nabla u|_{L^4}^4 \leq |\nabla u|_{L^\infty}^2|A^{1/2}u|^2 \leq c_2^2R^2|Au|^2(1+\log\lambda_n).$$

Thus, if $R \geq 1$ and R satisfies (18.16a), relation (16.17) will be satisfied provided

$$\frac{\lambda_{n+1}-\lambda_n}{2} > \frac{3}{\lambda_{n+1}}\frac{4}{(1-\delta)^2|Au|^4}(c_{12}R^{2p+2}(1+\log\lambda_n)^{p+1}|Au|^2 + |Ag|^2).$$

So assuming

$$R \geq \frac{4|Ag|}{1-\delta} + 1, \tag{18.18}$$

we conclude that (18.17) will be satisfied, and hence properties (I) and (II) in Chapter 9 too, if

$$\lambda_{n+1} - \lambda_n \geq \frac{(1-\log\lambda_n)^{n+1}}{\lambda_{n+1}}c_{13}\frac{R^{2p}}{(1-\delta)^2}. \tag{18.18a}$$

Property (IV), $\Gamma \subset C_{n,X}$ (see Chapter 9), is satisfied if $R > 4|g|(1-\delta)^{-1}$, thus, in particular, if (18.18) is valid. Finally, property (IV), $N(u)\mathbb{R} + T_u(\Gamma) \subset$

$C_{n,\gamma}$ (see Chapter 9), follows from (18.17) if

$$\lambda_{n+1} \geq \frac{1}{2\gamma^2}(\lambda_{n+1} - \lambda_n). \tag{18.18b}$$

Indeed, for this purpose (18.17) is stronger than (9.10), which implies (V), as noted in Chapter 9.

We therefore proved Proposition 18.1.

Proposition 18.1. *If* R *satisfies* (18.18) *and* λ_n, λ_{n+1} *satisfy* (18.18a), (18.18b), *then the ellipsoid*

$$\Gamma = \{u \mid P_n u = u, |A^{1/2}u| = R\}$$

satisfies condition (C1) *in Chapter* 10.

Continuing, we will check conditions (C2) to (C5) of Chapter 10. The set Y (considered in (C2) in Chapter 10 and introduced in Chapter 5) will be chosen as follows:

$$Y = \{u \mid |A^{1/2}u| \leq \beta_1(4|Ag|(1-\delta)^{-1} + 1),$$
$$|Au| \leq \beta_2(1 + |Ag|)^{1/2}c_8(q)(1-\delta)^{-(1+r)/2}\},$$

where β_1, β_2 are two parameters to be fixed later. Let R and n be as in Proposition 18.1 above and let $u_0 \in \Gamma$. In the argument following (18.8) we can take $R_0 = R$. We deduce

$$|A^{1/2}S(t)u_0| \leq R, \qquad t \geq 0. \tag{18.19}$$

Moreover, since $|Au_0| \leq \lambda_n^{1/2}R$, we deduce from (18.11)

$$|AS(t)u_0| \leq R_2 = \max\{\lambda_n^{1/2}R, c_8(q)R^{r/2}\}, \qquad t \geq 0. \tag{18.20}$$

Also from (18.10) and the remark following (18.13) we infer

$$|S(t)u_0|_{L^\infty} \leq c_2 R(1 + \log R_2)^{1/2} \leq c_{14}R(1 + \log \lambda_n + \log R)^{1/2}. \tag{18.21}$$

By taking

$$\beta_1 = R\left(\frac{4|Ag|}{1-\delta} + 1\right)^{-1}, \qquad \beta_2 = \lambda_n^{1/2}R(1-\delta)^{(1+p)/2}c_8(q)^{-1}(1 + |A^{1/2}g|)^{-r/2}$$

we have $\Gamma \subset Y$ and the inequalities (19.18) and (19.20) remain valid also for $u_0 \in Y$. The estimate

$$|L(t)v|^2 \leq k_1|v|^2 + k_2|A^{1/4}v|^2 + k_3|A^{1/2}v|^2 \quad \text{for } u_0 \in Y$$

is valid with $k_2 = k_3 = 0$ and any

$$k_1 \geq \sup\{|f'(S(t)u_0|_{L^\infty}^2 \mid u_0 \in Y, t \geq 0\};$$

thus we can take

$$k_1 = c_1^2[1 + c_{14}^p R^p(1 + \log \lambda_n + \log R)^{p/2}]^2$$

and therefore condition (C4) (in Chapter 10) is satisfied if

$$\frac{\lambda_{n+1} - \lambda_n}{2} > c_1(1 + c_{14}^p R^p(1 + \log \lambda_n + \log R)^{p/2}). \tag{18.22}$$

Notice that (18.22) also implies the analog of the inequality (4.6). The verification of (C3) is straightforward if one notices that (5.5) is true in our case with $k_9' = 0$ and that (5.6) has the same form as (18.22) with possible different absolute constants. In the same manner (16.22) also implies condition (C5). Therefore in (16.22) we shall take for c_1 and c_{14} the largest values requested by the above argument. Finally, using $R \geq 1, |Ag| \geq |g|$, and (18.8), (18.11) we infer that Y is also absorbing if

$$\lambda_n^{1/2} > c_8(q)(1 + |Ag|)^{r/2}(1 - \delta)^{-(1+r)/2}. \tag{18.23}$$

Summarizing, for R fixed satisfying (18.18) and for n satisfying (18.18a), (18.18b), (18.22), and (18.23), all the conditions (C1) to (C5) in Chapter 10 hold for the ellipsoid Γ (considered in Proposition 18.1).

It is only now that the constraint $p < 2$ plays a crucial role. Namely, it was proved in [R] that if $\{\Lambda_n\}_{n=1}^{\infty}$ denotes the increasing sequence of the distinct eigenvalues of A, then there are a subsequence $\{\Lambda_{m_k}\}_{k=1}^{\infty}$ and an absolute constant c_{15} such that

$$\Lambda_{m+1} - \Lambda_m \geq c_{15} \log \Lambda_m \quad \text{for all } m = m_k,\ k = 1, 2, \ldots. \tag{18.24}$$

It is now obvious that if λ_n is chosen equal to some Λ_{m_k} and is large enough, then all inequalities (18.18a), (18.18b), (18.22), and (18.23) are satisfied. We can therefore conclude with the following:

Theorem 18.2. *For all $\lambda_{n+1} > \lambda_n \in \{\Lambda_{m_j}\}_{j=1}^{\infty}$ large enough, there exists an inertial manifold of dimension n that is the closure of the graph of a function*

$$\Phi = \left\{u \in P_n H : |A^{1/2}u| \leq \frac{4|Ag|}{1-\delta} + 1\right\} \to H,$$

satisfying all the properties in Theorem 10.1.

Remark 18.3. The supplementary conditions one needs to impose in order to ensure the asymptotic completeness of the inertial manifold $\bar{\Sigma}$ (Theorem 12.1) are very simple. Indeed, k_9' and k_{14} can be taken to be 0 and thus all we need is to make sure that n is larger than some absolute constant.

Remark 18.4. One can relax the assumption (18.4), $f'(u) \leq \delta < 1$, requiring it to be valid only for large $|u|$. This would allow for more complicated attractors. Also one can consider u and f vector-valued and obtain the same results. Our choice was motivated by the desire to give a simple two-dimensional parabolic example in which the maximum principle is not used.

CHAPTER 19

Application: The Chaffee–Infante Reaction–Diffusion Equation

As an example of a parabolic reaction–diffusion equation with less stringent conditions than in Chapter 18, we briefly outline the construction of an inertial manifold for the Chaffee–Infante equation [H] in two dimensions:

$$\begin{cases} \dfrac{\partial u}{\partial t} - \Delta u + \lambda(u^3 - u) = 0, \\ \lambda > 0, \qquad \Omega = [-\pi, +\pi]^2 = T^2, \quad \text{periodic boundary conditions}, \\ u(0) = u_0 \end{cases} \tag{19.1}$$

(we do not restrict ourselves to odd periodic functions). For $\lambda > 1$, this equation admits multiple nonconstant steady states besides $u = 0$ and $u = \pm 1$. As it possesses a Lyapunov functional

$$V(t) = \frac{1}{2} \int_\Omega (\nabla u)^2 \, dx + \lambda \int_\Omega \left(\frac{u^4}{4} - \frac{u^2}{2} \right) dx,$$

the universal attractor consists of fixed points and their unstable sets.

Generically (a Baire set of λ's), the set of fixed points is finite, and the unstable sets are smooth manifolds. The global dynamics of (19.1) have been extensively investigated in [H] for $\Omega = [-\pi, +\pi]$. We closely follow the notation and proof of Chapter 18; the main interest lies in (i) relaxing the growth conditions on f', f''; (ii) using the maximum principle to simplify the construction of Chapter 18, avoiding absorbing sets in $H^2(T^2)$; and (iii) extending the construction of an integral inertial manifold to a situation where A is *not coercive* and only positive semidefinite. This requires a new definition of Γ and nontrivial adjustments to Chapter 9.

Denote by $\lambda_0 = 0$ the nontrivial zero eigenvalue of A and by $\lambda_1 = 1$ the first nonzero eigenvalue. It follows that

$$|A^{q+r}u| \geq |A^q u|, \qquad u \in H^1(T^2) \cap H^{q+r}(T^2), \quad \forall q > 0, \forall r \geq 0. \quad (19.2)$$

We recall:

Lemma 19.1 [H]. *Let $u(t)$ be a trajectory of* (19.1) *in $H^1(T^2)$. If $\|u(t_0)\|_{L^\infty} \leq 1 + \eta$ for some $t_0 \geq 0$, some $\eta > 0$, then for all $t \geq t_0$:*

$$\|u(t)\|_{L^\infty} \leq 1 + \eta.$$

PROOF. Apply Stampacchia's weak maximum principle in $H^1(T^2)$ to (19.1).

In constructing the appropriate absorbing set Y, we prove a much stronger:

Lemma 19.2. *For $\varepsilon > 0$ arbitrary, let*

$$Y = \{u \in H^1(T^2) \mid |u| \leq \rho_0 \quad \text{and} \quad |\nabla u| \leq \rho_1\},$$

where

$$\rho_0 = ((1 + \varepsilon)|\Omega|)^{1/2}, \qquad \rho_1 = (\lambda(1 + 2\varepsilon)|\Omega|)^{1/2} \quad (19.3)$$

Then Y is invariant under $\{S(t)\}_{t \geq 0}$ and absorbs all sets $Z \subset H^1(T^2)$ in a finite time $T(\varepsilon)$ independent of diam(Z). *Moreover, for all $t \geq T(\varepsilon) + 1$, we have $\|u(t)\|_{L^\infty} \leq \rho_\infty$, where*

$$\rho_\infty = C_1(\lambda + \tfrac{1}{2})^{1/4}(\rho_0 \rho_1)^{1/2} + 2\pi\rho_0, \quad (19.4)$$

C_1 being the constant for Agmon's inequality in T^2 for functions with zero mean.

Remark 19.3. However arbitrarily large u_0 is in $H^1(T^2)$, $u(t)$ enters the invariant (from Lemma 19.1) ball $\{u \in H^1 : \|u\|_{L^\infty} \leq \rho_\infty\}$ in a finite time uniform with respect to u_0.

PROOF OF LEMMA 19.2. It hinges on properties of nonlinear Gronwall's inequalities of the type

$$\frac{da}{dt} + Ca^2 - a \leq 0, \qquad a > 0, C > 0; \quad (19.5)$$

straightforward integration of the latter establishes that any ball of radius $1/C + \varepsilon$ is not only invariant but also absorbing in a finite time $T(\varepsilon)$ independent of $a(0)$. Now:

$$\frac{1}{2}\frac{\partial}{\partial t}|u|^2 + |\nabla u|^2 + \lambda\left[\frac{1}{|\Omega|}\left(\int_\Omega u^2\right)^2 - |u|^2\right] \leq 0; \quad (19.6)$$

hence there exists $T_0(\varepsilon)$, uniform with respect to u_0, such that the ball

$\{u : |u| \leq \rho_0\}$ is absorbing in $T_0(\varepsilon)$ and invariant. Moreover,

$$\frac{1}{2}\frac{\partial}{\partial t}|\nabla u|^2 + \frac{|\nabla u|^4}{|u|^2} - \lambda|\nabla u|^2 \leq 0, \tag{19.7}$$

where we used standard interpolations and the positivity of $3\lambda \int_\Omega u^2 |\nabla u|^2\, dx$. It follows that there exists $T_1(\varepsilon)$ such that $S(t)\{u : |u| \leq \rho_0\} \subset \{u : |\nabla u| \leq \rho_1\}$ for all $t \geq T_1(\varepsilon)$. Define $T(\varepsilon) = T_0(\varepsilon) + T_1(\varepsilon)$. Finally, let $t \in [T(\varepsilon), T(\varepsilon) + 1]$; from

$$\begin{aligned} \tfrac{1}{2}|\nabla u(T+1)|^2 + \int_T^{T+1} |\Delta u|^2\, dt &\leq \tfrac{1}{2}|\nabla u(T)|^2 + \lambda \int_T^{T+1} |\nabla u|^2\, dt \\ &\leq (\lambda + \tfrac{1}{2})\rho_1^2, \end{aligned} \tag{19.8}$$

we infer that there exists $T^* \in [T(\varepsilon), T(\varepsilon) + 1]$ such that $|Au(T^*)|^2 \leq (\lambda + \frac{1}{2})\rho_1^2$. The proof is completed with Agmon's inequality and Lemma 19.1. □

Corollary 19.4. *Let Z be an arbitrary bounded set in $H^1(T^2)$. Let $\theta = \max\{1, \sup_{u\in Z}|u|/\rho_0, \sup_{u \in Z}|\nabla u|/\rho_1\}$. Then $\{S(t)Z\}_{t\geq 0} \subset \theta Y$; moreover, θY is absorbing in a time $T(\theta)$ uniform with respect to u_0. Finally, $\|S(t)(\theta Y)\|_{L^\infty} \leq \theta\rho_\infty$, for all $t \geq 1$.*

Lemma 19.5. *The set*

$$Y \cap \{u \in H^2(T^2) : |Au| < \rho_2\},$$

where

$$\rho_2 = \sqrt{2\lambda\rho_1(1 + 3\rho_\infty^2)}, \tag{19.9}$$

is invariant under $\{S(t)\}_{t\geq 0}$ and absorbs all $Z \subset H^2(T^2)$ in a finite time $T_2(\varepsilon)$ independent of diam(Z) *in H^2.*

Proof. Let $t \geq T(\varepsilon) + 1$; then

$$\frac{1}{2}\frac{\partial}{\partial t}|Au|^2 + |A^{3/2}u|^2 \leq \lambda|Au|^2 + 3\lambda\|u\|_{L^\infty}^2|A^{3/2}u||A^{1/2}u|,$$

hence

$$\frac{\partial}{\partial t}|Au|^2 + \frac{|Au|^4}{|A^{1/2}u|^2} - 2\lambda^2|A^{1/2}u|^2\{1 + 9\|u\|_{L^\infty}^4\} \leq 0,$$

where we used the interpolation $|Au|^2 \leq |A^{1/2}u||A^{3/2}u|$. The proof is concluded by an argument similar to that used in the proof of Lemma 19.2. □

We now proceed to the construction of the inertial manifold. First, we must address the noncoercivity of A. Write (19.1) as $\partial u/\partial t + Au + R(u) = 0$, where $R(u) = \lambda(u^3 - u)$ is the nonlinearity. Restricting the dynamics to θY (i.e.,

$u(0) \in \theta Y$) ensures that $\|u(t)\|_{L^\infty} \le \theta\rho_\infty$ for $t \ge 0$ (cf. end of proof of Lemma 19.2). It follows that the strong coercivity hypothesis (4.5) is not needed to prove the strong squeezing property (Chapter 4) for the Chaffee–Infante equation; this property follows from the estimate

$$\frac{1}{2}\frac{d}{dt}|w|^2 + \{\Lambda(t) - k_7\}|w|^2 \le 0, \tag{19.10}$$

with k_7 and $\Lambda(t)$ defined as in Chapter 4, but setting $k_7 = \lambda(3\theta^2\rho_\infty^2 + 1)$, $k_4 = 1$. This replaces conditions (4.5), (4.5a), (4.5b) in Chapter 4. The same modifications carry over to Chapter 7 (cf. (7.2)).

In Chapter 5 (cone invariance properties), the estimate (5.5) is replaced by

$$|(L(t)w, w)| \le k_7|w|^2, \tag{19.11}$$

k_7 defined as in (19.10).

We define the projections P as the spectral projection on the span of the eigenspaces associated to $\lambda_0, \lambda_1, \ldots, \lambda_m$ (including the zero eigenvalue). One key ingredient in the proofs of Chapter 5 is the inequality

$$|A^{1/2}p|^2 \le \lambda_m|p|^2, \qquad P_m p = p,$$

which is certainly true for our A and our choice of P_m. Chapter 5 carries over and we are now ready to check and/or adopt the conditions (C1) to (C5) of Chapter 10. The key difference lies in the choice of initial data for our inertial manifold:

$$\Gamma = \{u \,|\, P_n u = u, |A^{1/2}u|^2 + \lambda|u|^2 = R^2\}, \tag{19.12}$$

with R and n large, to be determined later. If $u \in \Gamma$, it follows that

$$|Au| \le \lambda_n^{1/3} R. \tag{19.13}$$

The outward unnormalized normal to Γ at u is

$$\nu(u) = Au + \lambda u. \tag{19.14}$$

This renders the verification of conditions (I) to (V) in Proposition 9.2 somewhat more complex than in Chapter 18. We have

Lemma 19.6. *Let $u \in \Gamma$ and*

$$R^2 > 4\lambda|\Omega|; \tag{19.15a}$$

then

$$(N(u), \nu(u)) \ge \frac{|Au|^2}{2} + \lambda^2|u|^2. \tag{19.15b}$$

PROOF. Recalling the definition (19.14) of $\nu(u)$ we have

$$(N(u), \nu(u)) = |Au|^2 + 3\lambda\int_\Omega |\nabla u|^2 u^2\,dx + \lambda^2\left(\int_\Omega u^4\,dx - |u|^2\right).$$

A sufficient condition for (19.15b) to hold is

$$\tfrac{1}{2}|Au|^2 + \lambda^2\left(\frac{1}{|\Omega|}|u|^4 - 2|u|^2\right) \geq 0. \tag{19.16a}$$

This obviously holds if $|u|^2 > 2|\Omega|$. Thus we assume now that

$$|u|^2 \leq 2|\Omega|. \tag{19.16b}$$

But if R satisfies (19.15a) and $u \in \Gamma$

$$16\lambda^2|\Omega|^2 \leq |A^{1/2}u|^4 < |Au|^2 2|\Omega|;$$

hence $|Au|^2 \geq 8\lambda^2|\Omega|$, and this together with (19.16b) implies (19.16a). □

Choosing R as in (19.15a) we proceed to establish

Lemma 19.7. *Let*

$$\lambda_{n+1}(\lambda_{n+1} - \lambda_n) > \lambda C_2(\Omega)R^4(1 + \log\lambda_n)^2.$$

Then conditions (I) *and* (II) *of Proposition* 9.2 *are satisfied.*

PROOF. From (9.12), a sufficient condition is that for $u \in \Gamma$ we have

$$\frac{|Au + \lambda u|^2 |A^{1/2}(I - P_n)N(u)|^2}{(N(u), Au + \lambda u)^2} < \lambda_{n+1} - \lambda_n; \tag{19.17}$$

from (19.15b), this is certainly true if

$$\frac{8}{\lambda_{n+1}|Au|^2}|A(1 - P_n)N(u)|^2 < \lambda_{n+1} - \lambda_n. \tag{19.18}$$

Now

$$|A(1 - P_n)N(u)| \leq \lambda\|3u^2\|_{L^\infty}|Au| + 6\lambda\|u\|_{L^\infty}\left(\int_\Omega |\nabla u|\,dx\right)^{1/2}. \tag{19.19}$$

We shall use the following form of the first estimate (18.10):

$$\|u\|_{L^\infty} \leq m + c_4(\Omega)|A^{1/2}\hat{u}|\left\{1 + \left[\log\frac{|A\hat{u}|}{|A^{1/2}\hat{u}|}\right]^{1/2}\right\}, \qquad u \in H^2(T^2), \tag{19.20}$$

with m and $\hat{u}$ given by $m = |\int_\Omega u|/|\Omega|$ and $\hat{u} = u - (1/|\Omega|)\int_\Omega u\,dx$.
For $u \in \Gamma$, we have

$$\|u\|_{L^\infty} \leq R\sqrt{|\Omega|} + c_3(\Omega)R\{1 + (\log\lambda_n)^{1/2}\}, \tag{19.21a}$$

since $P_n u = u$; equivalently,

$$\|u\|_{L^\infty} \leq c_4(\Omega)R\{1 + (\log\lambda_n)^{1/2}\}. \tag{19.21b}$$

Injecting (19.21b) into (19.19), we have

$$\|3u^2\|_{L^\infty} \le c_5(\Omega)R^2\{1 + (\log\lambda_n)^{1/2}\}^2;$$

similarly,

$$36\|u\|_{L^\infty}^2 \le c_6(\Omega)R^2(1 + (\log\lambda_n)^{1/2})^2.$$

Finally,

$$\int_\Omega |\nabla u|^4\,dr \le R^2\|\nabla u\|_{L^\infty}^2 \le c_7(\Omega)R^2|Au|^2\{1 + (\log\lambda_n)^{1/2}\}^2,$$

where we noticed that $\int_\Omega \nabla u \equiv 0$ and applied (18.10). Collecting the above estimates in (19.19), we obtain a sufficient condition for (19.17) to hold:

$$\frac{\lambda}{\lambda_{n+1}}c_8(\Omega)R^4\{1 + (\log\lambda_n)^{1/2}\}^4 \le \lambda_{n+1} - \lambda_n, \tag{19.22}$$

which can be given the form in the statement of the lemma. □

Condition (V) in Chapter 9 now follows from (19.17) if $\lambda_{n+1} \ge (2\gamma)^{-2}(\lambda_{n+1} - \lambda_n)$.

We now complete the construction of the inertial manifold as in Chapter 10. From Lemma 19.6, Corollary 19.4, the construction of an invariant absorbing set θY such that $\{S(t)\Gamma\}_{t\ge 0} \subset \theta Y$ is immediate. We need a uniform estimate for $\|S(t)\Gamma\|_{L^\infty}$, $t \ge 0$, in order to adjust the construction of Chapter 10. This follows immediately from (19.21b) and Lemma 19.1:

$$\|S(t)\Gamma\|_{L^\infty} \le c_9(|\Omega|)R\{1 + (\log\lambda_n)^{1/2}\}. \tag{19.23}$$

Moreover, on $S(t)\Gamma$, the estimate $|L(t)v|^2 \le k_1|v|^2 + k_2|A^{1/4}v|^2 + k_3|A^{1/2}v|^2$ is valid with $k_2 = k_3 = 0$ and $k_1 \ge \sup \lambda^2\|(3u^2 - 1)^2\|_{L^\infty}$, $u \in S(t)\Gamma$. We can take

$$k_1 = c_{10}(\Omega)\lambda^2R^4\{1 + (\log\lambda_n)^{1/2}\}^4. \tag{19.24}$$

Remark 19.8. We also need L^∞ estimates for $S(t)u_0$, $t \ge 1$, $u_0 \in \theta Y$, but $u_0 \notin \Gamma$. From Corollary 19.4 it follows that $\|S(t)u_0\|_{L^\infty} \le \theta\rho_\infty$, for all $u_0 \in \theta Y$ and $t \ge 1$. In estimates (3.7), (3.8), we can set $k_2 = k_3 = 0$ and $k_{10} = c_{10}(\Omega)\lambda^2\theta^4\rho_\infty^4$. Now set

$$\theta = \max\left(1, \frac{R}{\rho_1}, \frac{R}{\rho_0\sqrt{\lambda}}\right) = c_{11}(\Omega)R,$$

hence

$$k_1 = c_{11}(\Omega)R^4(1 + \lambda^2)^2, \tag{19.25}$$

using the definition (19.4) for θ_∞. Hence $|L(t)v|^2 \le k_1|v|^2$, for all $u_0 \in \theta Y$, $t \ge 1$.

Combining (19.24) and (19.25), we obtain

$$\begin{cases} |L(t)v|^2 \le k_1'|v|^2, \\ k_1' = c_{12}(\Omega)(1+\lambda^2)^2R^4\{1+(\log\lambda_n)^{1/2}\}^4 \end{cases} \tag{19.26}$$

for all $t \ge 0$, and all $t \ge 1$ in θY.

We still need (19.26) for $u \in \{S(t)\Gamma\}_{t\ge 0}$, to ensure the injectivity of P_m everywhere on the inertial manifold, as opposed to the weaker condition (19.25). The above construction essentially bypasses the need for an absorbing set in $H^2(T^2)$ and makes full use of the weak maximum principle for the Chaffee–Infante equation.

We now make explicit the spectral gap condition that ensures condition (C4) in Chapter 10:

$$\frac{\lambda_{n+1}-\lambda_n}{2} > (k_1')^{1/2};$$

indeed, by (19.26) this will follow from

$$\frac{\lambda_{n+1}-\lambda_n}{2} > c_{12}(\Omega)^{1/2}(1+\lambda^2)R^2\{1+(\log\lambda_n)^{1/2}\}^2. \tag{19.27}$$

As in Chapter 18, if Λ_m, $m = 1, 2, \ldots$, denote the distinct eigenvalues of A, $\Lambda_1 = \lambda_1 < \Lambda_2 < \cdots$, we have a sequence of m's such that

$$\Lambda_{m_j+1} - \Lambda_{m_j} \ge c_3(\Omega)\Lambda_{m_j}. \tag{19.28}$$

As in Chapter 18, we can now conclude with the following:

Theorem 19.9. *For all $\lambda_{n+1} > \lambda_n \in \{\Lambda_{m_j}\}_{j=1}^{\infty}$ large enough, there exists an inertial manifold of dimension n that is the closure of the graph of a function*

$$\Phi = \{u \in P_nH : |A^{1/2}u|^2 + \lambda|u|^2 \le R^2\} \mapsto H$$

satisfying all properties in Theorem 10.1. *Here R is subjected to the conditions $R \ge 1$ and* (19.15a).

Remark 19.10. The above construction of an inertial integral manifold can easily be extended to the more general equation:

$$\begin{cases} \dfrac{\partial u}{\partial t} - \Delta u + \lambda f(u) = 0, \\ \Omega = T^2, \end{cases} \tag{19.29}$$

where $f(s) \in C^2(\mathbb{R})$ satisfies the following conditions:

$$\begin{cases} \text{(i) There exists } K_1 > 0 \text{ such that } f(s) > 0, \text{ for all } s \text{ such that } |s| \ge K_1; \\ \text{(ii) } \underline{\lim}\, f'(s) \ge -K_2 \text{ for some } K_2 \ge 0; \\ \text{(iii) for all } s,\ |f''(s)| \le K_3|s|^p + K_4 \text{ for some } K_3, K_4 \ge 0 \text{ and } p > 0. \end{cases} \tag{19.30}$$

There is only one difference from the Chaffee–Infante example: the absorption times $T_0(\varepsilon)$, $T_1(\varepsilon)$, $T_2(\varepsilon)$, as defined in Lemmas 19.2 and 19.5, are no longer uniform with respect to arbitrary initial data. This does not alter the arguments, and we leave this more general case for the reader.

Remark 19.11. Of course, Theorem 10.1 can be completed with the conclusion of Theorem 12.1 just by taking in Theorem 19.9, the inertial manifold of large enough dimension.

References

[Ag] Agmon, S., *Lectures on Elliptic Boundary Value Problems*, Elsevier, New York, 1965.

[BV] Babin, A. V., Vishik, M. I., Regular attractors of semigroups and evolution equations, *J. Math. Pures Appl.*, 62 (1983), 441–491.

[BV1] Babin, A. V., Vishik, M. I., *Usp. Mat. Nauk*, 38, 4 (232) (1983), 133–187.

[BPV] Berger, P., Pomeau, Y., Vidal, C., *L'Ordre dans le chaos*, Hermann, Paris, 1984.

[BLMcLO] Bishop, A. R., Forest, M. G., McLaughlin, D. W., Overman, E. A., A quasi-periodic route to chaos in a near-integrable PDE, *Physica D*, 23 (1986), 293–328.

[Ca] Carr, J., *Applications of Centre Manifold Theory*, Springer-Verlag, New York, 1981.

[ChH] Chow, S.-N., Hale, J. K., *Methods of Bifurcation Theory*, Springer-Verlag, New York, 1982.

[CoE] Collet, P., Eckman, J. P., *Iterated Maps of the Interval as Dynamical Systems*, Birkhauser, Boston, 1980.

[CF] Constantin, P., Foias, C., Sur le transport des variétés de dimension finie par les solutions des équations de Navier–Stokes, *C. R. Acad Sci. Paris*, 296, I (1983), 23–26.

[CF1] Constantin, P., Foias, C., Global Lyapunov exponents, Kaplan–Yorke formulas and the dimension of the attractor for 2D Navier–Stokes equations, *Comm. Pure Appl. Math.*, 38 (1985), 1–27.

[CFT] Constantin, P., Foias, C., Temam, R., Attractors representing turbulent flows, *Mem. Amer. Math. Soc.*, No. 314 (1985), 53.

[CFMT] Constantin, P., Foias, C., Manley, O. P., Temam, R., Determining modes and fractal dimension of turbulent flows, *J. Fluid Mech.*, 150 (1985), 427–440.

[CFNT] Constantin, P., Foias, C., Nicolaenko, B., Temam, R., *C. R. Acad. Sci. Paris*, 382, I (1986), 375–378.

[CHS] Conway, E., Hoff, D., Smoller, J., Large time behavior of solutions of nonlinear reaction–diffusion equations, *SIAM J. Appl. Math.*, 35, 11 (1978), 1–16.

[De] Devaney, R. L., *An Introduction to Chaotic Dynamical Systems*, Benjamin/Cummings, Menlo Park, Calif., 1986.

[DO] Douady, A., Oesterlé, T., Dimension de Hausdorff des attracteurs, *C. R. Acad. Sci. Paris*, 190, A (1980), 1135–1138.

[DS] Dunford, N., Schwartz, J. T., *Linear Operators*, Interscience, New York, 1957.

[FS] Foias, C., Saut, J-C., Asymptotic behavior as $t \to +\infty$ of solutions of Navier–Stokes equations and nonlinear spectral manifolds, *Indiana Univ. Math. J.*, 33 (1984), 459–471.

[FS1] Foias, C., Saut, J.-C., On the smoothness of the nonlinear spectral manifolds of Navier–Stokes equations, *Indiana Univ. Math. J.*, 33 (1984), 911–926.

[FS2] Foias, C., Saut, J.-C., Variété invariante a décroissance exponentielle lente pour les équations de Navier–Stokes avec forces potentielles, *C. R. Acad. Sci. Paris*, 302, Série I (1986), 563–566.

[FT] Foias, C., Temam, R., Some analytic and geometric properties of the solutions of the evolution Navier–Stokes equations, *J. Math. Pures Appl.*, 48 (1979), 339–368.

[FST] Foias, C., Sell, G. R., Temam, R., Variétés inertielles des équations différentielles dissipatives, *C. R. Acad. Sci. Paris*, 301, I (1985), 139–141.

[FST1] Foias, C., Sell, G. R., Temam, R., Inertial manifolds for nonlinear evolutionary equations, *J. Differential Equations*, in press.

[FNST] Foias, C., Nicolaenko, B., Sell, G. R., Temam, R., Variétés inertielles pour l'équations de Kuramoto–Sivashinsky, *C. R. Acad. Sci. Paris*, 301, I (1985), 285–288.

[FNST1] Foias, C., Nicolaenko, B., Sell, G. R., Temam, R., Inertial manifolds for the Kuramoto–Sivashinsky equations and an estimate of their lowest dimension, *J. Math. Pures Appl.*, in press.

[GH] Guckenheimer, J., Holmes, P., *Nonlinear Oscillations, Dynamical Systems, and Bifurcations of Vector Fields*, Springer-Verlag, New York, 1984.

[H] Haken, H., *Synergetics*, Springer, New York, 1982.

[HMO] Hale, J. K., Magalhaes, L. T., Oliva, W. M., *An Introduction to Infinite Dimensional Dynamical Systems—Geometric Theory*, Appl. Math. Sci., No. 47, Springer, New York, 1984.

[Hl] Hale, J., *Asymptotic Behavior of Dissipative Systems*, Mathematical Surveys and Monographs, Vol. 25, AMS, Providence, RI, 1988.

[HS] Hale, J. K., Sell, G., unpublished.

[Ha] Hartman, P., *Ordinary Differential Equations*, Birkhauser, Boston, 1982.

[He] Henry, D., *Geometric Theory of Parabolic Equations*, Lecture Notes in Math., No. 840, Springer-Verlag, New York, 1983.

[HN] Hyman, J. M., Nicolaenko, B., The Kuramoto–Sivashinsky equations, a bridge between PDEs and dynamical systems, *Physica D*, 18 (1986), 113–126.

[HN1] Hyman, J. M., Nicolaenko, B., Coherence and chaos in the Kuramoto–Velarte equation, in *Recent Developments in Nonlinear PDEs*, edited by M. G. Crandall and P. Rabinowitz, Academic Press, Orlando, Fla., 1987.

[HNZ] Hyman, J. M., Nicolaenko, B., Zaleski, S., Order and complexity in the Kuramoto–Sivashinsky model of turbulent interfaces, *Physica D*, 23 (1986), 265–292.

[M-P] Mallet-Paret, J., Negatively invariant sets of compact maps and an extension of a theorem of Cartwright, *J. Differential Equations*, 22 (1976), 331–348.

[M-PS] Mallet-Paret, J., Sell, G. R., Inertial manifolds for reaction-diffusion equations in higher space dimensions, IMA Preprint Series, No. 331, University of Minnesota, Minneapolis, 1987.

[M] Mañe, R., Reduction of semilinear parabolic equations to finite dimensional C^1 flows, *Lecture Notes in Math.*, No. 597, Springer-Verlag, New York, (1977), 361–378.

[MeP] de Melo, W., Palis, J., *Geometric Theory of Dynamical Systems*, Springer-Verlag, New York, 1982.

[Met] Métivier, G., Valeurs propres des operateurs définis par la restriction de systèmes variationelles a des sous-espaces, *J. Math. Pures Appl.*, 57 (1978), 133–156.

[NSh] Nicolaenko, B., Scheurer, B., Remarks on the Kuramoto–Sivashinsky equation, *Physica D*, 12 (1984), 391–395.

[NST] Nicolaenko, B., Scheurer, B., Temam, R., Quelques propriétes des attracteurs pour l'équation de Kuramoto–Sivashinsky, *C. R. Acad. Sci. Paris*, 298, I (1984), 23–25.

[NST1] Nicolaenko, B., Scheurer, B., Temam, R., Some global dynamical properties of the Kuramoto–Sivashinsky equations; nonlinear stability and attractors, *Physica D*, 16 (1985), 155–183.

[R] Richards, J., On the gaps between numbers which are the sum of two squares, *Adv. Math.*, 46 (1982), 1–2.

[SS] Sacker, R. J., Sell, G. R., A spectral theory for linear differential systems, *J. Differential Equations*, 27 (1978), 320–358.

[Sch] Schuster, H. J., *Deterministic Chaos; an Introduction*, Physik, Weinheim, 1984.

[T] Temam, R., *Infinite Dimensional Dynamical Systems in Mechanics and Physics*, Applied Mathematics Series, Vol. 68, Springer-Verlag, New York, 1988.

Index

Applied Mathematical Sciences

cont. from page ii

55. *Yosida:* Operational Calculus: A Theory of Hyperfunctions.
56. *Chang/Howes:* Nonlinear Singular Perturbation Phenomena: Theory and Applications.
57. *Reinhardt:* Analysis of Approximation Methods for Differential and Integral Equations.
58. *Dwoyer/Hussaini/Voigt (eds.):* Theoretical Approaches to Turbulence.
59. *Sanders/Verhulst:* Averaging Methods in Nonlinear Dynamical Systems.
60. *Ghil/Childress:* Topics in Geophysical Dynamics: Atmospheric Dynamics, Dynamo Theory and Climate Dynamics.
61. *Sattinger/Weaver:* Lie Groups and Algebras with Applications to Physics, Geometry, and Mechanics.
62. *LaSalle:* The Stability and Control of Discrete Processes.
63. *Grasman:* Asymptotic Methods of Relaxation Oscillations and Applications.
64. *Hsu:* Cell-to-Cell Mapping: A Method of Global Analysis for Nonlinear Systems.
65. *Rand/Armbruster:* Perturbation Methods, Bifurcation Theory and Computer Algebra.
66. *Hlaváček/Haslinger/Nečas/Lovíšek:* Solution of Variational Inequalities in Mechanics.
67. *Cercignani:* The Boltzmann Equation and Its Applications.
68. *Temam:* Infinite Dimensional Dynamical System in Mechanics and Physics.
69. *Golubitsky/Stewart/Schaeffer:* Singularities and Groups in Bifurcation Theory, Vol. II.
70. *Constantin/Foias/Nicolaenko/Temam:* Integral Manifolds and Inertial Manifolds for Dissipative Partial Differential Equations.
71. *Catlin:* Estimation, Control, and the Discrete Kalman Filter.
72. *Lochak/Meunier:* Multiphase Averaging for Classical Systems.
73. *Wiggins:* Global Bifurcations and Chaos.
74. *Mawhin/Willem:* Critical Point Theory and Hamiltonian Systems.
75. *Abraham/Marsden/Ratiu:* Manifolds, Tensor Analysis, and Applications, 2nd ed.
76. *Lagerstrom:* Matched Asymptotic Expansions: Ideas and Techniques.
77. *Aldous:* Probability Approximations via the Poisson Clumping Heuristic.
78. *Dacorogna:* Direct Methods in the Calculus of Variations.